PLUMBER'S EXAM
PREPARATION GUIDE

By

Howard C. Massey

Craftsman Book Company
6058 Corte del Cedro, P.O. Box 6500
Carlsbad, California 92008

Acknowledgements

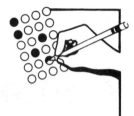

The author expresses his sincere thanks and appreciation to the following groups for granting permission to use direct or paraphrased quotes from their publications to authenticate the correct Code responses as formatted in this book:

International Association of Plumbing and Mechanical Officials

5032 Alhambra Avenue
Los Angeles, California 90032

National Fire Protection Association, Inc.

Batterymarch Park
Quincy, Massachusetts 02269

Southern Building Code Congress International, Inc.

900 Montclair Road
Birmingham, Alabama 35213

Portions of the *Standard Plumbing Code* and *Standard Gas Code* have been reprinted. The reprinted material is not necessarily complete, nor is it the official position of the Southern Building Code Congress International.

To my wife, Hilda,

for her encouragement and participation
that helped to make this book possible,
and
to my son, Richard, who is carrying on the
family plumbing tradition.

Library of Congress Cataloging in Publication Data

Massey, Howard C.
    Plumber's exam preparation guide.

    Includes index.
    1. Plumbing--Examinations, questions, etc.
I. Title.
TH6128.M37   1985   696'.1'076   85-19050
ISBN 0-934041-04-0

Contents

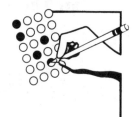

Introduction — How to Use This Book

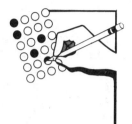

I hope you've picked up this book because you're looking for a good-paying career as a licensed journeyman or master plumber. That's exactly my goal: to launch your career by helping you get licensed. This book has the information you need to pass a plumbing exam based on either of the two popular national codes.

If you've been installing plumbing systems for years as an unlicensed plumber, this book is for you. There's no need to work under the handicap of not having a license. The information between the covers of this book will cover every subject that's likely to be on most plumbing exams.

If you're just starting as an apprentice plumber, this book is also for you. It begins at the beginning. You'll have no trouble understanding what's explained here. Read carefully and you'll soon earn the recognition that licensed professionals are entitled to in modern society.

In most communities, any plumber working *without supervision* must be licensed. Many states now require the certification of journeyman plumbers as well as specialty plumbers. This trend is sure to continue as legislatures recognize the need to protect the public from charlatans and the incompetent.

Let me issue a caution right at the beginning. Don't let anyone convince you that studying for a plumbing exam is a waste of time. It isn't. Most licensing authorities prepare demanding exams that are a good test of the examinee's knowledge. These exams guarantee that plumbing installed in modern buildings will meet minimum standards for protecting the lives and health of building occupants for many years.

If you don't believe that slipshod plumbing and haphazard sanitary systems can be a major health menace, you haven't traveled in foreign countries where plumbers are neither licensed nor held to reasonable standards of competence.

Begin your study for the exam with two points in mind. First, you're going to take the exam seriously. You'll pass, but only if you study carefully each of the questions and answers in this book. Second, every minute you spend studying this book is a minute well spent. What you learn for the exam is the foundation on which your professional career will be built.

Understand also that the licensing authority isn't the enemy. They aren't trying to keep you out of the plumbing profession. They only want to set some basic standards. The public should be assured that all licensed plumbers are knowledgeable professionals. That's good for society in general, and it's good for all professional plumbers who live and work in your community.

Before I go any farther, let me offer some information on my background. I've been an apprentice, journeyman and master plumber. For 15 years I ran my own plumbing contracting company. For 14 years I was assistant plumbing chief and plans examiner for a building department. I've helped write, monitor and grade plumber's exams. I have a pretty good idea of what you need to know to pass the exam.

Unfortunately, I see far too many applicants who are not well prepared when they sit down to take the test. Let me make this clear — taking the test without doing a good job of preparation is a complete waste of time — both yours and that of the licensing authority. The results are predictable. Don't make that mistake.

The most common reason for failure is that the applicant didn't study properly because he didn't know how, or studied the wrong material. This book should forever put an end to that excuse. You have in your hands the most complete, easiest-to-use, most practical reference available for preparing to take the tests that are actually given today. Read this book carefully, examine every question, understand all the answers. Do this, and there's no way you'll be unprepared on examination day.

All the common questions and answers are here, of course. But just knowing the answer isn't always enough. Sometimes it's just as important to understand *why* a particular answer is correct. That's why many answers include a quotation from the appropriate code reference. Sometimes the correct answer depends on which code is being used in the jurisdiction. If that's the case, I've given the correct answer for each of the two popular national codes. And sometimes you'll find notes or clarifications under the answer when there's an important point you might miss.

What to Expect

There was a time when a few years of experience and some knowledge of the gas and local plumbing code were almost enough to guarantee a passing grade. The old tests were usually closed book exams. No reference materials were permitted in the examination room. These tests evaluated the applicant's memory of the code and his ability to illustrate and design plumbing systems. That wasn't necessarily the best way to test a plumber's knowledge. No plumber has to work completely without reference books. Memorizing code sections isn't practical. It's also important that you know where to find an answer and have the background to interpret what the reference book says.

Today, you'll probably take an open book exam which asks you to solve practical problems and answer questions from recommended references. That's closer to the type of problems plumbers face every day in their work. Speed in locating the right reference for each question (and making the correct interpretation) is essential.

Most questions given on exams are based on the local plumbing and gas codes. Other test questions will likely be taken from references recommended by the examining authority. You'll probably receive a list of approved references when you apply to take the exam. These approved references are the only books allowed in the examination room.

The following is a typical list of approved references for a journeyman plumber's exam. But this is an *example only*. Make sure you use the actual list recommended by *your* testing authority.

Your local plumbing code, plus any applicable ordinances and amendments.
NFPA Pamphlet No. 54, Gas Appliances and Gas Piping
NFPA Pamphlet No. 14, Standpipe and Hose Systems
Plumbing, by H.E. Babbitt
Plumber and Pipe Fitters Library

Mathematics for Plumbers and Pipefitters
Plumbing 1, by Harry Slater
Related Information Plumbing 2, by Harry Slater
Blueprint Reading for Plumbers, Residential and Commercial
Plumbing Installation and Design
Student Guide for Plumbing Installation and Design

The master's exam list will be longer and includes several subjects that aren't covered in the references listed above.

Getting the Right Books

Get all the recommended references as soon as possible. If you live within driving distance of a well-stocked technical bookstore, they'll probably have most or all of what you need. Smaller general bookstores usually don't stock many technical books. But they may have *some* of the listed titles. Most bookstores are willing to special order books for you, but you'll have to wait four to six weeks for them to arrive.

Remember that books and pamphlets used to improve or maintain your professional skills are deductible on your income tax return. They're also valuable references even after you've passed the exam. Don't be afraid to spend what's needed to get the recommended books. They'll be a good investment.

Codes and Standards

Five major plumbing codes are used in the United States: *Basic Plumbing Code, ICBO Plumbing Code, Standard Plumbing Code, National Standard Plumbing Code* and the *Uniform Plumbing Code*. Several states have written their own plumbing codes. The five model codes are written by private organizations that have some interest in improving standards in the plumbing industry. By themselves, these model codes are not the law. They're written in hopes that some city, county or state will adopt them as a regulation. When your city, state or county does adopt a model code, it becomes the authority for all plumbing work done in that jurisdiction.

Of course, the code adopted is entirely up to the governing authority in your city, county or state. And that branch of government is free to amend, delete, or supplement the code that's actually adopted — and many do.

Almost all plumbing codes in the United States are "referral codes." They refer to other standard references when describing materials and design procedures. For example, every model plumbing code includes a table which lists all the plumbing materials acceptable for use within the jurisdiction. The *Standard Plumbing Code* states, "Plumbing fixtures shall be constructed from approved materials, have smooth impervious surfaces, be free from defects and concealed fouling surfaces, and shall conform to the standards listed in Table 500." The standards for plumbing fixtures as listed in Table 500 were developed by the *American National Standards Institute, Inc.* (ANSI).

You'll see many references like that in your plumbing code. The *Standard Plumbing Code* lists 31 separate standards in the plumbing section alone. A few of these references are ANSI (mentioned above), ASTM (American Society for Testing and Materials), CISPI (Cast Iron Soil Pipe Institute), FS (Federal Specifications) and NBS (National Bureau of Standards). All references in your code place a burden on you, the plumber, to understand what's required and comply with what's called for.

Questions in the plumbing systems section of this book are based on the two most popular national codes, the *Standard Plumbing Code* and the *Uniform Plumbing Code*. If you compare the code references for each question, you'll see how similar these plumbing codes actually are. In cases where there are some differences (mainly in the area where fixture units regulate pipe sizes and lengths), I've provided notes to explain the differences.

Most states adopt all or nearly all of one of these two popular codes and, of course, use that code as the authority for the state plumbing exam.

The *Standard Plumbing Code* is used in Alabama, Arkansas, Florida, Georgia, Louisiana, Mississippi, North Carolina, South Carolina, Tennessee, and some parts of Delaware, Missouri, Oklahoma, Texas, and West Virginia.

The *Uniform Plumbing Code* is used in Alaska, California, Hawaii, Idaho, Maine, Montana, Nevada, New Hampshire, Oregon, Utah, Washington, and some areas of Arizona, Colorado, Iowa, Kansas, Missouri, Nebraska, North Dakota, Oklahoma, Pennsylvania, South Dakota, Texas, West Virginia, and Wyoming.

If you don't live in one of the 34 states listed above, the answers to some questions may vary slightly from the answers given in this book. But the differences between most plumbing codes is growing smaller and smaller with each passing year. After all, what's good plumbing practice in Massachusetts is also good plumbing practice in Indiana.

In the section on gas systems, I've based the questions and answers on the *Standard Gas Code*. It's compatible with the popular *National Fuel Gas Code*, and probably with whatever gas code is adopted in your area. The *Standard Gas Code* provides (as do all gas codes) the minimum requirements for gas installations.

Here's an important point: All exam questions are based on *minimum code requirements*. If the minimum pipe size permitted under the code is 1/2" and you answer 3/4" just to play it safe, your answer is *incorrect*.

How to Prepare for the Exam

This book is a guide to preparing for the journeyman or master plumbing exam. It isn't a substitute for studying the recommended references and it won't teach you the plumbing trade. But it will give you a *complete knowledge of the type of questions* asked in the plumbing exam. It will also give you a "feel" for the examination and provide some of the confidence you need to pass.

Emphasis is on multiple-choice questions because that's what nearly all tests have now. I've grouped the questions into chapters. Each chapter covers a single subject. This will help you discover your strengths and weaknesses. Analyze the questions you miss on the practice exam at the back of this book. You'll probably notice you're weaker in some subjects than others. If you've missed a lot of the gas questions or many of the math questions, go back and study these areas again.

> *Most question and answer books for plumbers provide the correct answers at the end of each chapter or at the end of the book without explanation. This book has the correct answer after each question, based on each of the two most popular plumbing codes or the Standard Gas Code.*
>
> *When reading a question, cover the answer with a sheet of paper or card cut to size. Read the question carefully. Mark your answer on a separate sheet of paper before moving the paper that covers the correct answer. Then slide the paper down and check to see if your answer is correct. If it isn't, read the code responses to find out why it's wrong.*

How to Study

Set aside a definite time to study, following a schedule that meets your needs. Study two or three nights each week or all day on Saturdays. Study alone most of the time. But spend a day reviewing with a plumbing buddy before exam day. You can help each other dig out the facts and concepts you'll need to pass the exam.

Study in a quiet, well-lighted room that's respected as your study space by family members and friends. If it's hard to find a spot like that in your home, go to the neighborhood library where others are reading and studying.

Before you begin to study, spend a few minutes getting into the right frame of mind. That's important. You don't have to be an Einstein to pass the plumber's exam. But good motivation will nearly guarantee your success. No one can provide that motivation but *you*. Getting your license is a goal you set for yourself; it's your key to a satisfying career and a better paying job.

As you study each reference, highlight or underscore important points with a yellow marker or red felt tip pen. That makes it easier to find important passages when you're doing the final review — and when you're taking the test.

Put paper tabs on the corners of each major section in all the references you'll take into the exam room. On the portion of the tab that extends beyond the edge of the book, write the name of the section or the subject. That makes locating each section easier and quicker — an important consideration on an open book test. Speed in locating answers is important. In the sample exam in this book, which is based on actual exams, you'll have less than four minutes to answer each question!

Your study plan should allow enough time to review each reference at least three times. Read carefully the first time. The next review should take only about 10% of the time that the first reading took. Make a final review of all references and notes on the day before the exam. *This is the key to success in passing the exam: Review, review, review!* The more you review, the better your grasp of the information and the faster you'll be able to find the answers.

The Examination

Your examination questions were probably compiled from lists submitted by members of the plumber's examination board. Board members usually include several senior plumbing contrac-

tors, perhaps a college professor, a registered engineer, and a code authority like a plumbing plans examiner. The exam will include *code, practical,* and *theoretical* questions. Some boards prefer theoretical questions. Others favor practical and code questions. No matter which type your examining authority emphasizes, this book will help you get prepared.

In areas where the journeyman or master plumbing exam is given two or three times each year, the examining authority will have several basic exams that are used in rotation. But the same examination will never be administered twice in a row.

The test writers maintain a bank of several hundred questions covering each test subject. Questions are selected at random, and chances are that some of the questions on any exam have already been used on an earlier examination.

Many questions are known as *universal truths.* With minor variations, these questions will be on nearly every plumber's exam in the country. This book is filled with the questions that pop up on nearly every plumbing exam.

Although plumbing is a complex trade, it's encouraging to note that there are only so many subject areas that any test can cover. And many of the questions on the exam will closely resemble questions in this book.

Types of Questions

Nearly all examination questions will be objective. This means you won't be required to draw complex piping isometrics of DWV or water piping systems and you won't have to write any essays. But many examinations do require that you at least identify which isometrics are *wrong* and draw simple corrections.

One major examining board gave the following instructions to all plumbers taking their certification examination:

The afternoon portion of the examination (four hours in duration), given on the first day, *has been changed.* Although all of the 80 questions are related to codes, approximately 10 questions will concern the interpretation of *isometric* drawings in which the examinee will be required to identify errors in the drawings, if any, in accordance with code requirements. In addition, another 10 questions will require the examinee to examine isometric drawings. If the drawings are not in conformance with codes, the examinee will be required to redraw isometrics correctly in the spaces provided.

As you know, the lines on isometric drawings represent pipe and fittings. Symbols are used to show the location and type of fixtures. If your examining board requires reading and drawing of isometrics, you'll need additional preparation for the exam. *Plumbers Handbook,* by this author, explains how to read and create plumbing isometrics. If your local bookstore doesn't have *Plumbers Handbook,* use the order form at the back of this manual. Once you understand the key principles, it's easy to read and make isometric drawings.

The Answer Sheet

Following this introduction, you'll find a sample answer sheet that was used for a major plumbing examination. Answer sheets like these are designed for computer grading. Each question on the exam is numbered. Usually there will be four or five possible responses for each question. You'll be required to mark the best answer on the answer sheet.

Here's an example. The question is:

1) Atlanta is the capital city of the state of:

(A) Florida		(C) Arizona
(B) Texas		(D) Georgia

You should mark answer *D* for question 1 on the answer sheet.

Your answer sheet may vary slightly from the one that follows this section. But no matter what the answer sheet looks like, be sure to follow any instructions on that sheet! Putting the right answers in the wrong section will almost certainly cause you to fail.

Examination Day

On the day of your examination, listen to any oral instructions given and carefully read the printed directions. Failing to follow instructions will probably disqualify you.

There won't be any trick questions on most exams. Examination boards usually take their work very seriously. But the test writers will probably include at least a few questions that have to be read very carefully to be understood. The question may look familiar and the answer may seem obvious. But re-reading the question may point out some subtle distinction that makes the obvious answer totally wrong.

Any time the answer seems obvious at first glance, read the question again. Always look for the qualifying word or phrase in the question. Words like *always, never, least, most likely, smallest, but not less than, shall* and *may* can be dynamite. They can change the whole meaning of the question.

Sometimes several of the answers may seem possible. But only one will be correct. If you're not sure of the answer, use the process of elimination. Strike out answers you *know* are wrong. Then select the most likely of the answers that remain. This can change your odds from five-to-one to two-to-one on a question. Don't ever assume that there's an answer pattern. I've never seen a planned answer pattern on a plumbing exam. By chance, there may be a short series of answers that go "a, a, c, a, a, c, a, a. But don't assume that the next answer is "c". It probably isn't, and you'll probably miss several questions if you think you see a pattern in the answers and try to follow it. Read each question carefully and give the answer you think is correct.

Most important, pace yourself. Spend the first minute or two after the exam is passed out looking over the entire test booklet. Make an estimate of how many minutes should be allowed for each section or for each question. Check your progress after each 30 minutes. Most applicants won't finish all questions. Any question you don't answer will aways be wrong, of course. Time will nearly always be at a premium on an open book exam. With enough time everyone could get 100%! Using your time wisely may be half the battle.

Don't spend too much time on the toughest questions. It's a mistake to squander 10 minutes on the hardest question in the exam (and get it wrong) and then leave several relatively easy questions unanswered because you ran out of time. My advice is to skip the hard questions on the first pass. Then come back to them as time permits.

If you complete the exam early, don't leave the room. Spend the remaining time reviewing your answers. Try to find at least *one* error. It could mean the difference between passing or failing the examination. Many applicants do fail by just one point. Don't find yourself in that position. Make the most of every second available.

Organization of This Book

I've included here questions on gas systems, specialized plumbing systems and several other plumbing-related topics. There are two reasons for this. First, many exams include questions on these subjects. Second, this information is not readily available in the standard reference books. You may have trouble finding books that cover these questions.

This book is organized into five sections. Part One has questions and answers and code responses on plumbing systems. Part Two has questions and answers and code responses on gas systems. Part Three has questions and answers and code responses for more specialized plumbing subjects. Part Four has questions and answers and solutions (where applicable) on plumbing-related topics.

Part Five is a sample examination. Take this test two or three days before you are to take the actual exam. Use it to spot areas where you need extra review.

Let's Get Started

Enough of the preliminaries. It's time to get started with the questions and answers. Used correctly, this book will give you the confidence you need *now* to prepare thoroughly for the upcoming examination.

Happy studying! And best wishes.

Plumber's Examination Answer Sheet

Name _____
Please print   (last)   (first)   (middle)

Address _____

_____

Location of Examination_____

Signature _____

Directions For Marking Answer Sheet
- Use a black lead pencil only, #2 or softer. • DO NOT use ink or ballpoint pen.
- Make heavy black marks that fill the circle completely. • ERASE cleanly any answer you wish to change.
- Make NO stray marks on this answer sheet.

1 Ⓐ Ⓑ Ⓒ Ⓓ Ⓔ	26 Ⓐ Ⓑ Ⓒ Ⓓ Ⓔ	51 Ⓐ Ⓑ Ⓒ Ⓓ Ⓔ	76 Ⓐ Ⓑ Ⓒ Ⓓ Ⓔ
2 Ⓐ Ⓑ Ⓒ Ⓓ Ⓔ	27 Ⓐ Ⓑ Ⓒ Ⓓ Ⓔ	52 Ⓐ Ⓑ Ⓒ Ⓓ Ⓔ	77 Ⓐ Ⓑ Ⓒ Ⓓ Ⓔ
3 Ⓐ Ⓑ Ⓒ Ⓓ Ⓔ	28 Ⓐ Ⓑ Ⓒ Ⓓ Ⓔ	53 Ⓐ Ⓑ Ⓒ Ⓓ Ⓔ	78 Ⓐ Ⓑ Ⓒ Ⓓ Ⓔ
4 Ⓐ Ⓑ Ⓒ Ⓓ Ⓔ	29 Ⓐ Ⓑ Ⓒ Ⓓ Ⓔ	54 Ⓐ Ⓑ Ⓒ Ⓓ Ⓔ	79 Ⓐ Ⓑ Ⓒ Ⓓ Ⓔ
5 Ⓐ Ⓑ Ⓒ Ⓓ Ⓔ	30 Ⓐ Ⓑ Ⓒ Ⓓ Ⓔ	55 Ⓐ Ⓑ Ⓒ Ⓓ Ⓔ	80 Ⓐ Ⓑ Ⓒ Ⓓ Ⓔ
6 Ⓐ Ⓑ Ⓒ Ⓓ Ⓔ	31 Ⓐ Ⓑ Ⓒ Ⓓ Ⓔ	56 Ⓐ Ⓑ Ⓒ Ⓓ Ⓔ	81 Ⓐ Ⓑ Ⓒ Ⓓ Ⓔ
7 Ⓐ Ⓑ Ⓒ Ⓓ Ⓔ	32 Ⓐ Ⓑ Ⓒ Ⓓ Ⓔ	57 Ⓐ Ⓑ Ⓒ Ⓓ Ⓔ	82 Ⓐ Ⓑ Ⓒ Ⓓ Ⓔ
8 Ⓐ Ⓑ Ⓒ Ⓓ Ⓔ	33 Ⓐ Ⓑ Ⓒ Ⓓ Ⓔ	58 Ⓐ Ⓑ Ⓒ Ⓓ Ⓔ	83 Ⓐ Ⓑ Ⓒ Ⓓ Ⓔ
9 Ⓐ Ⓑ Ⓒ Ⓓ Ⓔ	34 Ⓐ Ⓑ Ⓒ Ⓓ Ⓔ	59 Ⓐ Ⓑ Ⓒ Ⓓ Ⓔ	84 Ⓐ Ⓑ Ⓒ Ⓓ Ⓔ
10 Ⓐ Ⓑ Ⓒ Ⓓ Ⓔ	35 Ⓐ Ⓑ Ⓒ Ⓓ Ⓔ	60 Ⓐ Ⓑ Ⓒ Ⓓ Ⓔ	85 Ⓐ Ⓑ Ⓒ Ⓓ Ⓔ
11 Ⓐ Ⓑ Ⓒ Ⓓ Ⓔ	36 Ⓐ Ⓑ Ⓒ Ⓓ Ⓔ	61 Ⓐ Ⓑ Ⓒ Ⓓ Ⓔ	86 Ⓐ Ⓑ Ⓒ Ⓓ Ⓔ
12 Ⓐ Ⓑ Ⓒ Ⓓ Ⓔ	37 Ⓐ Ⓑ Ⓒ Ⓓ Ⓔ	62 Ⓐ Ⓑ Ⓒ Ⓓ Ⓔ	87 Ⓐ Ⓑ Ⓒ Ⓓ Ⓔ
13 Ⓐ Ⓑ Ⓒ Ⓓ Ⓔ	38 Ⓐ Ⓑ Ⓒ Ⓓ Ⓔ	63 Ⓐ Ⓑ Ⓒ Ⓓ Ⓔ	88 Ⓐ Ⓑ Ⓒ Ⓓ Ⓔ
14 Ⓐ Ⓑ Ⓒ Ⓓ Ⓔ	39 Ⓐ Ⓑ Ⓒ Ⓓ Ⓔ	64 Ⓐ Ⓑ Ⓒ Ⓓ Ⓔ	89 Ⓐ Ⓑ Ⓒ Ⓓ Ⓔ
15 Ⓐ Ⓑ Ⓒ Ⓓ Ⓔ	40 Ⓐ Ⓑ Ⓒ Ⓓ Ⓔ	65 Ⓐ Ⓑ Ⓒ Ⓓ Ⓔ	90 Ⓐ Ⓑ Ⓒ Ⓓ Ⓔ
16 Ⓐ Ⓑ Ⓒ Ⓓ Ⓔ	41 Ⓐ Ⓑ Ⓒ Ⓓ Ⓔ	66 Ⓐ Ⓑ Ⓒ Ⓓ Ⓔ	91 Ⓐ Ⓑ Ⓒ Ⓓ Ⓔ
17 Ⓐ Ⓑ Ⓒ Ⓓ Ⓔ	42 Ⓐ Ⓑ Ⓒ Ⓓ Ⓔ	67 Ⓐ Ⓑ Ⓒ Ⓓ Ⓔ	92 Ⓐ Ⓑ Ⓒ Ⓓ Ⓔ
18 Ⓐ Ⓑ Ⓒ Ⓓ Ⓔ	43 Ⓐ Ⓑ Ⓒ Ⓓ Ⓔ	68 Ⓐ Ⓑ Ⓒ Ⓓ Ⓔ	93 Ⓐ Ⓑ Ⓒ Ⓓ Ⓔ
19 Ⓐ Ⓑ Ⓒ Ⓓ Ⓔ	44 Ⓐ Ⓑ Ⓒ Ⓓ Ⓔ	69 Ⓐ Ⓑ Ⓒ Ⓓ Ⓔ	94 Ⓐ Ⓑ Ⓒ Ⓓ Ⓔ
20 Ⓐ Ⓑ Ⓒ Ⓓ Ⓔ	45 Ⓐ Ⓑ Ⓒ Ⓓ Ⓔ	70 Ⓐ Ⓑ Ⓒ Ⓓ Ⓔ	95 Ⓐ Ⓑ Ⓒ Ⓓ Ⓔ
21 Ⓐ Ⓑ Ⓒ Ⓓ Ⓔ	46 Ⓐ Ⓑ Ⓒ Ⓓ Ⓔ	71 Ⓐ Ⓑ Ⓒ Ⓓ Ⓔ	96 Ⓐ Ⓑ Ⓒ Ⓓ Ⓔ
22 Ⓐ Ⓑ Ⓒ Ⓓ Ⓔ	47 Ⓐ Ⓑ Ⓒ Ⓓ Ⓔ	72 Ⓐ Ⓑ Ⓒ Ⓓ Ⓔ	97 Ⓐ Ⓑ Ⓒ Ⓓ Ⓔ
23 Ⓐ Ⓑ Ⓒ Ⓓ Ⓔ	48 Ⓐ Ⓑ Ⓒ Ⓓ Ⓔ	73 Ⓐ Ⓑ Ⓒ Ⓓ Ⓔ	98 Ⓐ Ⓑ Ⓒ Ⓓ Ⓔ
24 Ⓐ Ⓑ Ⓒ Ⓓ Ⓔ	49 Ⓐ Ⓑ Ⓒ Ⓓ Ⓔ	74 Ⓐ Ⓑ Ⓒ Ⓓ Ⓔ	99 Ⓐ Ⓑ Ⓒ Ⓓ Ⓔ
25 Ⓐ Ⓑ Ⓒ Ⓓ Ⓔ	50 Ⓐ Ⓑ Ⓒ Ⓓ Ⓔ	75 Ⓐ Ⓑ Ⓒ Ⓓ Ⓔ	100 Ⓐ Ⓑ Ⓒ Ⓓ Ⓔ

Sample answer sheet

Part One
Plumbing Systems

- **General Regulations**
- **Plumbing Definitions**
- **Materials: Quality and Weights**
- **Joints and Connections**
- **Traps and Cleanouts**
- **Sanitary Drainage Systems**
- **Vents and Venting**
- **Special Traps, Interceptors and Separators**
- **Indirect and Special Waste Piping**
- **Private Sewage Disposal Systems**
- **Water Distribution Systems**
- **Storm Water Drainage Systems**
- **Plumbing Fixtures and Special Plumbing Fixtures**

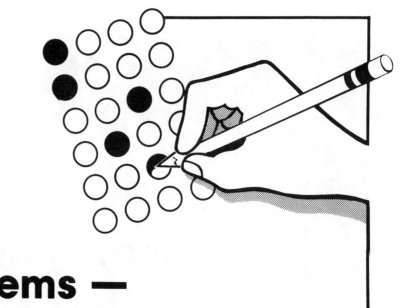

Chapter 1

Plumbing Systems — General Regulations

The general regulations in the plumbing code are intended as broad principles that apply to all plumbing work. For example, general regulations require the proper disposal of sewage and other waste materials.

General regulations cover (1) fittings used in direction changes, (2) fittings that are prohibited, (3) repairs and alterations to existing buildings, (4) trenching, excavation, and backfill, (5) structural safety, (6) protection of pipes, (7) location of plumbing fixtures, and much more. They require that the plumbing design, installation and workmanship conform with accepted engineering practices.

The questions in this chapter are a comprehensive test of these regulations for plumbing systems. Don't underestimate the importance of this part of the code. Every plumber needs a thorough understanding of the general regulations.

1-1 The Plumbing Code is used to:

(A) enforce the law
(B) confuse the plumber

(C) explain why plumbing installations may vary
(D) explain why minor changes from the basic rules are acceptable

___

Standard Plumbing Code

Uniform Plumbing Code

Answer: A
Code response: The provisions of the following chapters and sections shall constitute and be known and cited as the Code (law).

Answer: A
Code response: This ordinance shall be known as the Code (law).

Note: Plumbing codes are **laws** adopted by local and state authorities. Violating any provision of these codes is considered a misdemeanor, punishable by a fine or imprisonment upon conviction.

1-2 According to the Code, one of the following shall not be used in a drainage system:

(A) a 3-inch short sweep
(B) a fitting having 60-degree branches

(C) a double sanitary tee
(D) a straight tee branch

___

Standard Plumbing Code

Uniform Plumbing Code

Answer: D
Code response: A straight tee branch shall not be used as a drainage fitting

Answer: D
Code response: No tee branch shall be used as a drainage fitting.

1-3 According to the Code, the drainage system or any part thereof shall not be covered until:

(A) after a plumbing inspector has been called
(B) it has been tested

(C) it has been inspected
(D) it has been approved

___

Standard Plumbing Code

Uniform Plumbing Code

Answer: D
Code response: The plumbing system or any part thereof shall not be covered until approved.

Answer: D
Code response: No plumbing or drainage system or part thereof shall be covered until it has been accepted.

1-4 According to the Code, it shall be the duty and responsibility of the _____ to determine if the plumbing has been inspected before it is covered or concealed.

(A) general contractor
(B) permit holder

(C) plumbing inspector
(D) building and zoning record control branch

___

Standard Plumbing Code

Uniform Plumbing Code

Answer: B
Code response: Permit holder is responsible for all plumbing work installed.

Answer: B
Code response: Permit holder is responsible for work performed.

1-5 Every plumbing fixture directly connected to the drainage system, according to the Code, shall:

(A) have an overflow
(B) be maintained in a sanitary manner

(C) have hot and cold water
(D) be equipped with a water-seal trap

Standard Plumbing Code

Uniform Plumbing Code

Answer: D
Code response: Be equipped with a water-seal trap.

Answer: D
Code response: Be trapped by an approved type water-seal trap.

1-6 According to the Code, drainage systems water-tested for tightness and for inspection shall be tested with no less than a ____ head of water.

(A) 6-foot
(B) 8-foot

(C) 10-foot
(D) 12-foot

Standard Plumbing Code

Uniform Plumbing Code

Answer: C
Code response: No section shall be tested with less than a 10-foot head of water.

Answer: C
Code response: No section shall be tested with less than a 10-foot head of water.

1-7 According to the Code, when notching the ends of joists to install water piping, the depth of the notches shall not exceed ____ of the depth of the joists.

(A) 1/16
(B) 1/32

(C) 1/8
(D) 1/4

Standard Plumbing Code

Uniform Plumbing Code

Answer: D
Code response: When notching the ends of joists, the depth shall not exceed 1/4 the depth of the joists.

Answer: D
Code response: When notching, no structural member shall be seriously weakened or impaired.

Note: The Uniform Plumbing Code provides a warning but does not set a particular standard.

1-8 The Code requires that drainage fittings have threads that are tapped to allow for:

(A) full thread engagement
(B) grade

(C) simple adjustment
(D) pipe expansion

Standard Plumbing Code

Uniform Plumbing Code

Answer: B
Code response: None.

Answer: B
Code response: The threads of drainage fittings shall be tapped so as to allow for grade.

Note. The Standard Plumbing Code does not address this particular question. It is common knowledge, though, that manufacture of threaded drainage fittings allows for **grade** or **pitch**. Answer "B" is correct.

Q 1-9 The Code requires that when a water service pipe is installed in the same trench with a building sewer, there must be a minimum separation of _____ inches.

(A) 6
(B) 10

(C) 12
(D) 18

Standard Plumbing Code

Uniform Plumbing Code

Answer: C
Code response: The bottom of the water service pipe, at all points, shall be at least 12 inches above the top of the sewer pipe.

Answer: C
Code response: The bottom of water service pipe, at all points, shall be at least 12 inches above the top of the sewer pipe.

Q 1-10 The Code requires that sewage or other waste from a plumbing system be _____ before it is discharged into any waterway.

(A) discharged through a settling tank
(B) treated with Clorox

(C) rendered innocuous
(D) determined detrimental

Standard Plumbing Code

Uniform Plumbing Code

Answer: C
Code response: Sewage or other waste must first be rendered innocuous.

Answer: C
Code response: Sewage or other waste must first be rendered innocuous.

Q 1-11 According to the Plumbing Code, _____ fixture unit(s) flow rate shall be deemed to be 1 cubic foot per minute.

(A) 1
(B) 2

(C) 3
(D) 4

Standard Plumbing Code

Uniform Plumbing Code

Answer: A
Code response: A fixture unit flow rate shall be deemed to be 1 cubic foot of water per minute.

Answer: A
Code response: One fixture unit discharge capacity shall be deemed to be 7½ gallons per minute.

Q 1-12 According to the Code, 2-inch copper tubing installed horizontally must be supported at distances of not less than _____ feet.

(A) 4
(B) 6

(C) 8
(D) 10

Standard Plumbing Code

Uniform Plumbing Code

Answer: D
Code response: Copper tubing shall be supported at approximately 10-foot intervals for piping 2 inches and larger.

Answer: D
Code response: Copper tubing shall be supported at approximately 10-foot intervals for piping 2 inches and larger in diameter.

Q 1-13 A single-family house, connected to a public sewer, may discharge <u>all but one</u> of the following substances through its drainage system without violating the Code:

(A) human excrement

(B) potable water

(C) fireplace ashes

(D) clothes washer water

Standard Plumbing Code

Uniform Plumbing Code

Answer: C

Code response: No substance which will clog the pipes shall be allowed to enter the building drainage system.

Answer: C

Code response: It shall be unlawful for any person to deposit, into any plumbing fixture which is connected to any public sewer, any ashes.

Q 1-14 According to the Code, vertical _____ must be supported at the base and at each story level at intervals not exceeding 15 feet.

(A) cast-iron soil pipe

(B) threaded pipe

(C) copper tubing

(D) lead pipe

Standard Plumbing Code

Uniform Plumbing Code

Answer: A

Code response: Cast-iron soil pipe shall be supported at the base and at each story level at intervals not exceeding 15 feet.

Answer: A

Code response: Cast-iron soil pipe shall be supported at every story level or closer.

Q 1-15 All buildings intended for human habitation, according to Code, must:

(A) have an adequate supply of clear water

(B) be connected to a public sewer

(C) be maintained in a sanitary manner

(D) have an adequate supply of potable running water

Standard Plumbing Code

Uniform Plumbing Code

Answer: D

Code response: All buildings intended for human habitation shall be provided with an adequate, safe and potable water supply through a safe system of piping.

Answer: D

Code response: Each plumbing fixture shall be provided with an adequate supply of potable running water piped in an approved manner.

Q 1-16 According to Code, the ladder required by an inspector so that he can make a final inspection must be furnished by the:

(A) general contractor

(B) roofing contractor

(C) building and zoning department

(D) plumbing contractor

Standard Plumbing Code

Uniform Plumbing Code

Answer: D

Code response: Equipment necessary for the inspection shall be furnished by the plumber.

Answer: D

Code response: The equipment necessary for inspection shall be furnished by the person to whom the permit is issued.

Q 1-17 According to Code, pipe fittings that _____ shall not be used in sanitary drainage plumbing systems.

(A) are metallic
(B) are nonmetallic

(C) increase in size
(D) have abnormal flow obstruction

Standard Plumbing Code

Uniform Plumbing Code

Answer: D
Code response: Fittings that offer abnormal obstruction to flow shall not be used.

Answer: D
Code response: No fitting which obstructs the flow of sewage shall be used.

Q 1-18 A plumber requests an inspection and the work fails the test. According to Code, the plumber:

(A) has 10 days to make corrections

(B) shall make necessary corrections, then resubmit the work for inspection

(C) must make necessary corrections while the inspector waits

(D) may make the necessary corrections, then may resubmit the work for inspection

Standard Plumbing Code

Uniform Plumbing Code

Answer: B
Code response: If the official finds that the work will not pass the test, the permit holder **shall** be required to make necessary corrections, and the work **shall** then be resubmitted for inspection.

Answer: B
Code response: If the Administrative Authority finds that the work will not pass the test, necessary corrections **shall** be made, and the work **shall** then be resubmitted for test or inspection.

Note: The qualifying word in both Code responses is **shall**, a mandatory term.

Q 1-19 In existing buildings in which plumbing installations are to be renovated, the Code may permit the work to be done:

(A) only in a workmanlike manner
(B) with necessary deviations, providing the intent of the Code is met

(C) if like materials are used for replacement

(D) without a permit

Standard Plumbing Code

Uniform Plumbing Code

Answer: B
Code response: In existing buildings in which plumbing installations are to be renovated, deviations from the provisions of this Code may be permitted, providing such deviations conform to the intent of the Code.

Answer: B
Code response: In existing buildings in which plumbing installations are to be renovated, deviations from the provisions of this Code are permitted, provided such deviations are necessary.

Note: Both Codes, though wordy, eventually reach a similar conclusion. Answer "B" is correct.

Q 1-20 According to Code, the water service pipe supplying water for a two-bedroom, one-bath house must be:

(A) installed in 20-foot lengths when possible
(B) buried at least 12 inches below grade

(C) no less than ¾-inch I.D. piping
(D) no less than ¾-inch O.D. piping

Standard Plumbing Code

Uniform Plumbing Code

Answer: C
Code response: Water service lines shall in no case be less than ¾-inch nominal diameter.

Answer: C
Code response: No building water supply pipe shall be less than ¾-inch in diameter.

Q 1-21 According to Code, the horizontal distance "X" must be a minimum of _____ feet.

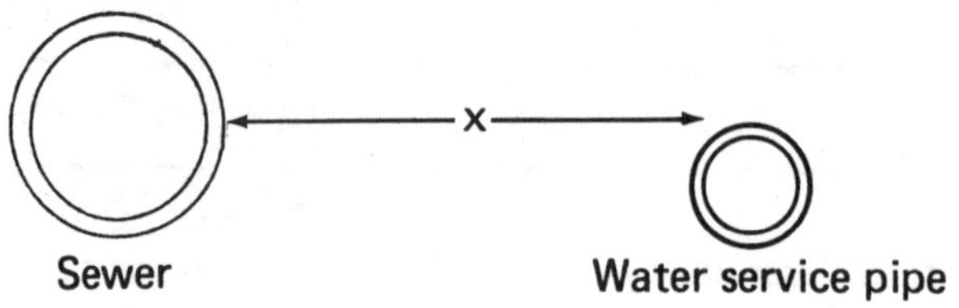

Figure 1-1

(A) 3
(B) 5

(C) 8
(D) 10

Standard Plumbing Code

Uniform Plumbing Code

Answer: B
Code response: Underground water service pipe and building sewer shall not be less than 5 feet apart horizontally.

Answer: _____
Code response: The Uniform Plumbing Code does not address the horizontal separation of sewer and water service pipe.

Note: For state exams using the Uniform Plumbing Code as their principal reference, this question would not appear. For journeyman exams, check local Code to determine if the question is addressed.

Q 1-22 In open-trench pipe installation, the Code requires a minimum of _____ inches of clean fill to be added above the top of piping before mechanical equipment is used for backfilling.

(A) 4
(B) 6

(C) 12
(D) 18

Standard Plumbing Code

Uniform Plumbing Code

Answer: C
Code response: Reasonably clean backfill shall be placed 12 inches over the pipe.

Answer: C
Code response: Trenches shall be backfilled in thin layers to 12 inches above top of piping with clean earth.

Q 1-23 In lieu of the water test, the Code will accept one of the following for a final test of a completed drainage and vent system:

(A) air
(B) gas mixture of oil of peppermint

(C) oxygen
(D) mercury vapor

Standard Plumbing Code

Uniform Plumbing Code

Answer: A
Code response: The air test is an acceptable method for testing drainage and vent systems.

Answer: A
Code response: Air testing a drainage and vent system is an acceptable method prescribed in this Code.

Q 1-24 According to Code, a single or double sanitary tee may be used in drainage lines where the direction of flow is from:

(A) horizontal to vertical
(B) horizontal to horizontal
(C) vertical to horizontal
(D) vertical to vertical

Standard Plumbing Code

Uniform Plumbing Code

Answer: A
Code response: In drainage lines, a single or double sanitary tee may be used when the direction of flow is from horizontal to vertical.

Answer: A
Code response: In drainage lines, a sanitary tapped tee may be used on a vertical line as a fixture connection.

Note: A fixture connection (flow) is from the horizontal to the vertical. The correct answer for both Codes is "A."

Q 1-25 One of the following fittings is <u>not</u> acceptable by Code for changes in direction in drainage piping:

(A) wye
(B) 1/5-bend
(C) long sweep ¼-bend
(D) 1/6-bend

Standard Plumbing Code

Uniform Plumbing Code

Answer: B
Code response: A 1/5-bend is not listed as an approved fitting for changes in direction in a drainage piping system.

Answer: B
Code response: A 1/5-bend is not listed as an approved fitting for changes in direction in a drainage piping system.

Q 1-26 Whenever compliance with all the provisions of the Code fails to eliminate _____, the owner or his agent has to make acceptable corrections.

(A) bad workmanship
(B) a drainage problem
(C) a nuisance
(D) uncertified workers from a job

Standard Plumbing Code

Uniform Plumbing Code

Answer: C
Code response: Whenever compliance with all the provisions of this Code fails to eliminate or alleviate a nuisance. . .

Answer: C
Code response: Whenever compliance with all the provisions of this Code fails to eliminate or alleviate a nuisance. . .

Q 1-27 According to Code, piping installed underground must be:

(A) installed a minimum of 6 inches below grade
(B) nonmetallic to prevent corrosion
(C) graded to drain to low point
(D) supported throughout its entire length

Standard Plumbing Code

Uniform Plumbing Code

Answer: D
Code response: Buried piping shall be supported throughout its entire length.

Answer: D
Code response: In-ground piping shall be laid on a firm bed throughout its entire length and shall be adequately supported.

Q 1-28 The plumber decides to water-test the drainage system in its entirety. According to Code, before inspection starts, the water must be kept in the system for a minimum of _____ minutes.

(A) 5
(B) 10

(C) 15
(D) 20

Standard Plumbing Code

Uniform Plumbing Code

Answer: C
Code response: The water shall be kept in the system for at least 15 minutes before inspection starts.

Answer: C
Code response: The water shall be kept in the system for at least 15 minutes before inspection starts.

Q 1-29 The Code specifies that all trenching required for the installation of a plumbing system within the walls of a building be:

(A) not closer than 5 feet to a building foundation
(B) not deeper than 3 feet maximum

(C) covered immediately after installation for safety of other workers
(D) open trench work

Standard Plumbing Code

Uniform Plumbing Code

Answer: D
Code response: All excavation for installation of a building drainage system shall be open trench work.

Answer: D
Code response: All excavation required to be made for the installation of a building drainage system shall be open trench work.

Q 1-30 According to Code, no plumbing permit can be considered valid until:

(A) plans are approved by plumbing official
(B) prescribed fees are paid

(C) architect's seal is imprinted upon approved plans
(D) permit is signed by qualifier

Standard Plumbing Code

Uniform Plumbing Code

Answer: B
Code response: No permit shall be valid until fees prescribed in this section shall have been paid.

Answer: B
Code response: Such applicant shall pay for each permit, at the time of issuance, a fee in accordance with the following schedule...

Q 1-31 Spaces between piping, sleeves, walls and floors, according to Code, must be:

(A) provided with a minimum annular space of ½-inch
(B) provided with a vermin-proof cap

(C) filled or tightly caulked
(D) filled to the top with fine sand

Standard Plumbing Code

Uniform Plumbing Code

Answer: C
Code response: Annular spaces between sleeves and pipes shall be filled or tightly caulked.

Answer: C
Code response: Voids around piping passing through masonry floors shall be appropriately sealed.

Q 1-32 According to Code, a 3-inch waste pipe passing through a foundation wall must be protected from external loadings or against differential settlement. A _____-inch cast-iron sleeve built into the wall is acceptable for this purpose.

(A) 3½

(B) 4

(C) 5

(D) 6

Standard Plumbing Code

Answer: B
Code response: A cast-iron sleeve two pipe sizes greater than the pipe passing through the wall is acceptable by Code.

Uniform Plumbing Code

Answer: _____
Code response: All piping passing under or through walls shall be protected from breakage.

Note: Since the next 2 pipe sizes are 3½ and 4, "B" would be the correct answer. The Standard Plumbing Code is specific in its requirements for the protection of piping passing through walls. Since the Uniform Plumbing Code leaves considerable discretion to the installer, this particular question would likely not appear on any exam where the Uniform Plumbing Code is the principal Code reference. For journeyman exams, check local Code to determine if above question is specifically addressed.

Q 1-33 According to Code, vent piping shall not be drilled or tapped:

(A) unless approved by the Administrative Authority

(B) for the purpose of making connections thereto

(C) unless drilled opening is to be used for roof-mounted A.C. condensate drain connection

(D) without prior approval by job architect

Standard Plumbing Code

Answer: A
Code response: Vent piping shall not be drilled or tapped, unless approved by the Administrative Authority.

Uniform Plumbing Code

Answer: B
Code response: Vent piping shall not be drilled or tapped for the purpose of making connections thereto.

Note: Each of the two Codes (if used as the principal Code reference for an exam) has a different correct answer. For journeyman exams, check local Code requirements for correct answer.

Q 1-34 According to Code, a fitting having a hub in the direction opposite to flow in the drainage system:

(A) must not be used

(B) may be used if the pipe is cut by a saw

(C) may be used by special permission of the plumbing inspector

(D) may be used if drainage piping is above grade

Standard Plumbing Code

Answer: B
Code response: Shall not be used, unless the pipe is cut by either a saw or snap cutter.

Uniform Plumbing Code

Answer: A
Code response: No double-hub fitting shall be used as a drainage fitting.

Note: Both Codes prohibit the use of this particular type of fitting. Using the above wording, there is a different correct answer for each Code.

Q **1-35** Using the illustration in Figure 1-2, according to Code, the installation:

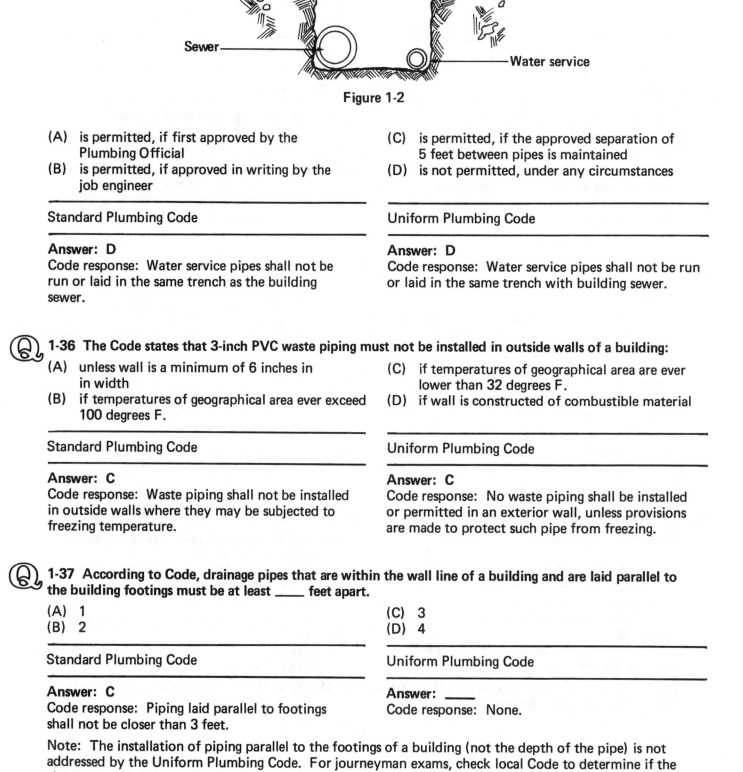

Figure 1-2

(A) is permitted, if first approved by the Plumbing Official

(B) is permitted, if approved in writing by the job engineer

(C) is permitted, if the approved separation of 5 feet between pipes is maintained

(D) is not permitted, under any circumstances

Standard Plumbing Code	Uniform Plumbing Code

Answer: D
Code response: Water service pipes shall not be run or laid in the same trench as the building sewer.

Answer: D
Code response: Water service pipes shall not be run or laid in the same trench with building sewer.

Q **1-36** The Code states that 3-inch PVC waste piping must not be installed in outside walls of a building:

(A) unless wall is a minimum of 6 inches in in width

(B) if temperatures of geographical area ever exceed 100 degrees F.

(C) if temperatures of geographical area are ever lower than 32 degrees F.

(D) if wall is constructed of combustible material

Standard Plumbing Code	Uniform Plumbing Code

Answer: C
Code response: Waste piping shall not be installed in outside walls where they may be subjected to freezing temperature.

Answer: C
Code response: No waste piping shall be installed or permitted in an exterior wall, unless provisions are made to protect such pipe from freezing.

Q **1-37** According to Code, drainage pipes that are within the wall line of a building and are laid parallel to the building footings must be at least _____ feet apart.

(A) 1

(B) 2

(C) 3

(D) 4

Standard Plumbing Code	Uniform Plumbing Code

Answer: C
Code response: Piping laid parallel to footings shall not be closer than 3 feet.

Answer: _____
Code response: None.

Note: The installation of piping parallel to the footings of a building (not the depth of the pipe) is not addressed by the Uniform Plumbing Code. For journeyman exams, check local Code to determine if the above question is specifically addressed.

Q 1-38 According to Code, vent pipe penetration of roof surfaces must be:

(A) a minimum of 7 inches above adjacent roof

(B) a minimum of 8 feet from nearest window

(C) provided with increasers to prevent frost closure

(D) made watertight

Standard Plumbing Code

Uniform Plumbing Code

Answer: D
Code response: Joints at the roof around vent pipes shall be made watertight.

Answer: D
Code response: Joints at the roof around vent pipes shall be made watertight.

Q 1-39 Plumbing fixtures, according to Code, cannot be located in such a manner:

(A) as to prevent their use by the physically handicapped

(B) as to prevent calculation of minimum fixture unit load in sizing drainage piping

(C) as to confuse the need for additional fixtures

(D) as to interfere with the normal operation of a door

Standard Plumbing Code

Uniform Plumbing Code

Answer: D
Code response: Fixtures shall not be located in such a manner as to interfere with the normal operation of doors.

Answer: D
Code response: Fixtures shall not be so located as to interfere with the normal operation of doors.

Q 1-40 Pipe trench excavation deeper than and parallel to footings, according to Code, must not be placed within an angle of pressure as transferred from the base of the structure to the sides of the excavation. The piping must be ＿＿ from the footings.

(A) 25 degrees

(B) 35 degrees

(C) 45 degrees

(D) 55 degrees

Standard Plumbing Code

Uniform Plumbing Code

Answer: C
Code response: Piping installed deeper than and parallel to footings shall be 45 degrees therefrom.

Answer: C
Code response: Trenches deeper than and parallel to the footings of any building must be 45 degrees therefrom.

Q 1-41 Having a unique plumbing installation, the plumber decides to test the system with air. According to Code, air must be forced into the system until there is a uniform gauge pressure of ＿＿ pounds per square inch.

(A) 5

(B) 10

(C) 15

(D) 20

Standard Plumbing Code

Uniform Plumbing Code

Answer: A
Code response: The air test shall be made by forcing air into the system until there is a uniform gauge pressure of 5 pounds per square inch.

Answer: A
Code response: The air test shall be made by forcing air into the system until there is a uniform gauge pressure of 5 pounds per square inch.

Q 1-42 The Code specifies that vertically installed piping must be secured at sufficiently close intervals to:

(A) carry the weight of the pipe
(B) carry the weight of the pipe and its contents

(C) prevent strains or stresses
(D) prevent sagging

Standard Plumbing Code

Uniform Plumbing Code

Answer: B
Code response: Vertical piping shall be secured at sufficiently close intervals to carry the weight of the pipe and contents.

Answer: B
Code response: Vertical piping shall be secured at sufficiently close intervals to carry the weight of the pipe and its contents.

Q 1-43 In accordance with Code, horizontally installed piping must be supported at sufficiently close intervals to:

(A) carry the weight of the pipe
(B) carry the weight of the pipe and its contents

(C) prevent strains or stresses
(D) prevent sagging

Standard Plumbing Code

Uniform Plumbing Code

Answer: D
Code response: Horizontal piping shall be supported at sufficiently close intervals to prevent sagging.

Answer: D
Code response: Horizontal piping shall be supported at sufficiently close intervals to prevent sagging.

Q 1-44 Code requires that all piping installed in corrosive type material be:

(A) first approved by the Plumbing Official
(B) supported properly

(C) protected properly
(D) in accordance with manufacturer's recommendations

Standard Plumbing Code

Uniform Plumbing Code

Answer: C
Code response: Pipes passing through or under corrosive material shall be protected against external corrosion.

Answer: C
Code response: All piping passing through or under corrosive material shall be protected in an approved manner.

Q 1-45 Code requires that vertical galvanized steel piping be supported at not less than:

(A) 8-foot intervals
(B) 10-foot intervals

(C) every story height
(D) every other story height

Standard Plumbing Code

Uniform Plumbing Code

Answer: D
Code response: Threaded pipe shall be supported at not less than every other story.

Answer: D
Code response: Screwed pipe shall be supported at not less than every other story height.

1-46 According to Code, bases of soil stacks must be supported:

(A) with concrete blocks
(B) by concrete piers

(C) to the satisfaction of the plumbing inspector
(D) to the satisfaction of the job engineer

Standard Plumbing Code	Uniform Plumbing Code
Answer: C Code response: Bases of all soil stacks shall be supported to the satisfaction of the plumbing official.	**Answer: C** Code response: All piping shall be adequately supported to the satisfaction of the Administrative Authority.

1-47 Codes requires that horizontally hung cast-iron soil pipe in 10-foot lengths have minimum spacing of supports not more than _____ feet apart.

(A) 5
(B) 8

(C) 10
(D) 12

Standard Plumbing Code	Uniform Plumbing Code
Answer: C Code response: Horizontal cast-iron soil pipe shall be supported at not more than 10-foot intervals on 10-foot lengths.	**Answer: C** Code response: Suspended cast-iron soil pipe that exceeds 5 feet in length may be supported at not more than 10-foot intervals.

1-48 According to Code, a water test may be applied to the drainage and vent system of a two-story residence:

(A) only in its entirety
(B) only in sections

(C) only during working hours
(D) in its entirety or in sections

Standard Plumbing Code	Uniform Plumbing Code
Answer: D Code response: The water test shall be applied to the drainage system in its entirety or in sections.	**Answer: D** Code response: The water test shall be applied to the drainage and vent system either in its entirety or in sections.

1-49 To meet the Code, all sections of a plumbing drainage and vent system must be tested with:

(A) water
(B) mercury vapor

(C) air or water
(D) air

Standard Plumbing Code	Uniform Plumbing Code
Answer: C Code response: All sections of a plumbing drainage and vent system must be tested with a water test or by an air test.	**Answer: C** Code response: All sections of a plumbing drainage and vent system must be tested with a water test or by an air test.

Q 1-50 The air test for a plumbing drainage and vent system must, by Code, be made by attaching the air compressor or other test apparatus to:

(A) the lowest fixture outlet available

(B) the highest vent opening

(C) any suitable opening

(D) a dishwasher waste opening

Standard Plumbing Code	Uniform Plumbing Code
Answer: C Code response: The air test for a plumbing drainage and vent system shall be made by attaching the air compressor or other test apparatus to any suitable opening.	**Answer: C** Code response: The air test for a plumbing drainage and vent system shall be made by attaching the air compressor or other test apparatus to any suitable opening.

Q 1-51 The Code specifies that a satisfactory hanger which supports any piping must be:

(A) of same material as piping

(B) accessible

(C) same size as pipe used

(D) of sufficient strength

Standard Plumbing Code	Uniform Plumbing Code
Answer: D Code response: A satisfactory hanger which supports any pipe must be of sufficient strength to maintain its proportionate share of the pipe alignment and to prevent sagging.	**Answer: D** Code response: A satisfactory hanger which supports any pipe must be of sufficient strength to maintain its proportionate share of the pipe alignment and to prevent sagging.

Q 1-52 The Code permits that a plumbing system installed before January 1, 1982, may be altered:

(A) without complying with latest Code requirements

(B) but, must comply with latest Code requirements

(C) without obtaining a permit

(D) and Code requirements waived if work is less than 50 percent of the existing system

Standard Plumbing Code	Uniform Plumbing Code
Answer: A Code response: In existing buildings in which plumbing installations are to be altered, necessary deviations from the provisions of this Code may be permitted.	**Answer: A** Code response: No provision of this Code shall be deemed to require a change in an existing building when such work was installed in accordance with the law in effect prior to the effective date of this Code.

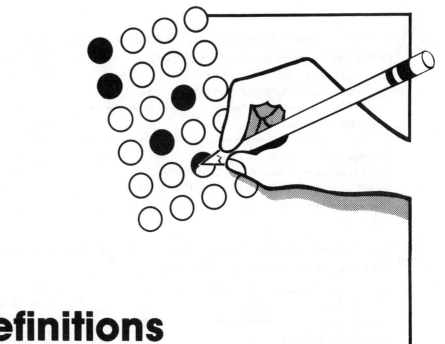

Chapter 2

Plumbing Definitions

You can't interpret and apply the plumbing code if you don't understand the words and terms it uses. Look up any of the general terms you're not familiar with in a good dictionary. There are some terms, however, which have taken on a special meaning because of their specialized use within the plumbing field. To avoid misunderstanding, there's a section of the code that defines the terms that have a special meaning when used in the code.

You'll also find these definitions in examinations for certification of journeyman plumbers and plumbing contractors.

The questions in this chapter will test your knowledge of these special words and their code definitions. State and local exams will question you extensively about the definitions. It's important that you know these special terms and be able to recognize them throughout the code.

2-1 A flushometer valve is a device, according to the Code, which is used to:

(A) flush the contents from a toilet bowl

(B) discharge a predetermined quantity of water to fixtures for flushing purposes

(C) automatically reseal floor drains

(D) flush all surfaces of a floor-mounted urinal

Standard Plumbing Code	Uniform Plumbing Code

Answer: B
Code response: Discharge a predetermined quantity of water to plumbing fixtures for flushing purposes.

Answer: B
Code response: Discharge a predetermined quantity of water to plumbing fixtures for flushing purposes.

 2-2 A horizontal waste and vent pipe which is sized to provide free circulation of air above the flow line of the drain pipe is defined by Code as a:

(A) combination waste and vent system

(B) continuous wet vent system

(C) nonsolvent waste and vent system

(D) ventilating piping system

Standard Plumbing Code	Uniform Plumbing Code

Answer: A
Code response: A combination waste and vent system is a special horizontal wet venting system using a common pipe to receive the waste from one or more sinks, floor drains, etc.

Answer: A
Code response: A combination waste and vent system is a specially designed system of waste piping embodying the horizontal wet vent of one or more sinks, floor drains, etc. by means of a common waste and vent pipe.

2-3 A "dead end," as defined by the Code is:

(A) a water main that makes a 90-degree turn and is capped

(B) a gas line that continues from horizontal to vertical for future use

(C) a branch extending a minimum of 2 feet beyond a building drain and plugged

(D) a utility alley used for piping installation that comes to an end

---

Standard Plumbing Code	Uniform Plumbing Code

Answer: C
Code response: A "dead end" is a branch leading from a DWV drainage system, terminating at a developed distance of 2 feet or more and plugged.

Answer: _____
Code response: "Dead end" is not defined.

Note: "Dead end" is not included in the definitions used in the Uniform Plumbing Code. Therefore, the term "dead end" would not be used in an examination where the Uniform Plumbing Code is the required Code reference.

Q 2-4 The Code defines a vertical vent that is a continuation of the drain to which it connects as a:

(A) common vent
(B) branch vent

(C) relief vent
(D) continuous vent

Standard Plumbing Code

Uniform Plumbing Code

Answer: D
Code response: A vertical vent that is a continuation of the drain to which it connects is a continuous vent.

Answer: D
Code response: A vertical vent that is a continuation of the drain to which it connects is a continuous vent.

Q 2-5 A vent pipe which receives waste from a shower is defined by Code as:

(A) a wet vent
(B) a common vent

(C) a circuit vent
(D) an island vent

Standard Plumbing Code

Uniform Plumbing Code

Answer: A
Code response: A wet vent is a pipe which receives waste from fixtures other than water closets.

Answer: A
Code response: A wet vent is a pipe which serves as a drain for other fixtures.

Q 2-6 As defined by Code, the primary purpose of a yoke vent is to:

(A) connect a waste stack to a vent stack
(B) connect a vent stack to a soil stack

(C) connect a loop vent to a stack vent
(D) prevent pressure changes in the stacks

Standard Plumbing Code

Uniform Plumbing Code

Answer: D
Code response: A yoke vent is used for the purpose of preventing pressure changes in the stacks.

Answer: D
Code response: A yoke vent is used for the purpose of preventing pressure changes in the stacks.

Q 2-7 The Code defines water that is safe for drinking purposes as:

(A) soft water
(B) potable water

(C) hard water
(D) clear water

Standard Plumbing Code

Uniform Plumbing Code

Answer: B
Code response: Potable water is satisfactory for drinking and meets the requirements of the Health Authority having jurisdiction.

Answer: B
Code response: Potable water is satisfactory for drinking and meets the requirements of the Health Authority having jurisdiction.

Q 2-8 The primary purpose of the Code in defining special terms is to:

(A) avoid misunderstanding

(B) include all possible words

(C) make them easier to find

(D) avoid duplication of words

Standard Plumbing Code

Uniform Plumbing Code

Answer: A

Code response: The primary purpose of defining terms in this Code is to avoid misunderstanding.

Answer: A

Code response: The primary purpose of defining terms in this Code is to avoid misunderstanding.

Q 2-9 The term "industrial wastes," as defined by the Code, would <u>not</u> include liquids containing:

(A) chemicals in solution

(B) chemicals in suspension

(C) well water

(D) fecal matter

Standard Plumbing Code

Uniform Plumbing Code

Answer: D

Code response: Industrial wastes resulting from the processes employed in industrial establishments are free of fecal matter.

Answer: D

Code response: Industrial waste means any and all liquid waste from industrial processes, except sewage.

Q 2-10 The term "liquid waste," as defined by Code, is waste that is discharged into a plumbing system from all but one of the following fixtures:

(A) bidet

(B) bed-pan washer

(C) wall-hung urinal

(D) cuspidor

Standard Plumbing Code

Uniform Plumbing Code

Answer: B

Code response: Liquid waste is the discharge from any fixture which does not receive fecal matter.

Answer: B

Code response: Liquid waste is the discharge from any fixture which does not receive fecal matter.

Q 2-11 A roof drain, as defined by Code, is a drain installed to receive water collecting on the surface of a roof, and to discharge the water into the:

(A) leader

(B) building storm drain

(C) combined building sewer

(D) public sewer

Standard Plumbing Code

Uniform Plumbing Code

Answer: A

Code response: A roof drain is a drain installed to receive water collecting on the surface of a roof and to discharge it into the leader.

Answer: _____

Code response: None.

Note: The Uniform Plumbing Code does not include "roof drain" in its range of definitions. Check local Code to determine if it's included in its definitions.

Q 2-12 Using Figure 2-1, pipe "A" is defined by Code as a:

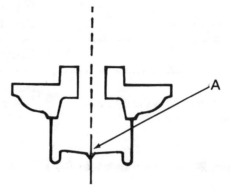

Figure 2-1

(A) unit vent
(B) revent

(C) double vent
(D) common vent

Standard Plumbing Code

Uniform Plumbing Code

Answer: D
Code response: A common vent is a vent at the junction of 2 fixture drains and serving as a vent for both fixtures.

Answer: D
Code response: 2 fixtures may be served by a common vent pipe when each such fixture empties wastes separately into an approved double fitting having inlet openings at the same level.

Q 2-13 Using Figure 2-2, pipe "A" is defined by Code as a:

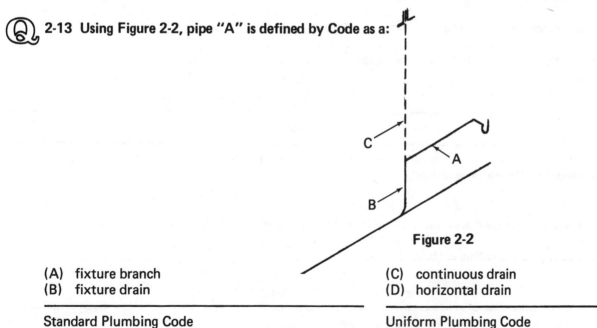

Figure 2-2

(A) fixture branch
(B) fixture drain

(C) continuous drain
(D) horizontal drain

Standard Plumbing Code

Uniform Plumbing Code

Answer: B
Code response: A fixture drain is the drain pipe from the trap of a fixture to the junction with any other drain pipe.

Answer: B
Code response: A fixture drain is the drain from the trap of a fixture to the junction with any other drain pipe.

Q 2-14 Again, referring to **Figure 2-2** on previous page, pipe "B" is defined by Code as a:

(A) wet pipe
(B) waste pipe

(C) vertical pipe
(D) continuous pipe

Standard Plumbing Code	Uniform Plumbing Code
Answer: B Code response: A waste pipe conveys only liquid waste, free of fecal matter.	**Answer: B** Code response: A waste pipe conveys only liquid waste, free of fecal matter.

Q 2-15 Refer again to **Figure 2-2** on previous page. Pipe "C" is defined by Code as a:

(A) vertical vent
(B) relief vent

(C) stack vent
(D) vent stack

Standard Plumbing Code	Uniform Plumbing Code
Answer: C Code response: A stack vent (dry portion) is the extension of a waste stack (as illustrated in Figure 2-2) above the highest horizontal drain connected to the stack.	**Answer: C** Code response: A stack vent (dry portion) is the extension of a waste stack (as illustrated in Figure 2-2) above the highest horizontal drain connected to the stack.

Practical interpretation and application of the Plumbing Code are possible only when you understand all of the words and terms used in the Code. The following 8 questions are presented to test your ability to correctly identify certain common words and terms used in the Plumbing Code.

Q 2-16 The Code **precludes** using a full "S" trap in a drainage system.

The word **precludes,** as used in this sentence, means:

(A) includes
(B) permits

(C) prevents
(D) allows

Universal Source: Dictionary

Answer: C
Dictionary response: To make impossible.

Q 2-17 The criteria for testing the plumbing system were established by the Code.

The word **criteria,** as used in this sentence, means:

(A) standards
(B) costs

(C) efforts
(D) usage

Universal Source: Dictionary

Answer: A
Dictionary response: A standard on which a judgment may be based. An established rule or principle for testing anything.

Q 2-18 The Administrative Authority may approve use of an alternate material, as long as the intention of the Code is met.

The word **intention,** as used in this sentence, means:

(A) wording
(B) objection
(C) impulse
(D) objective

Universal Source: Dictionary

Answer: D
Dictionary response: A determination to obtain a certain objective or goal.

Q 2-19 Any trap which has interior partitions is not recognized by the Code.

The word **recognized,** as used in this sentence, means:

(A) declared
(B) approved
(C) perceived
(D) demanded

Universal Source: Dictionary

Answer: B
Dictionary response: To admit as being of a particular status.

Q 2-20 Traps for floor drains must be provided with a means of access for cleaning.

The word **access,** as used in this sentence, means:

(A) permission
(B) a means of taking
(C) flexibility
(D) a means of approach

Universal Source: Dictionary

Answer: D
Dictionary response: A way or means of approach, admittance.

Q 2-21 The installation method for spacing lavatories is fairly uniform.

The word **uniform,** as used in this sentence, means:

(A) unchanging
(B) fast
(C) varying
(D) slow

Universal Source: Dictionary

Answer: A
Dictionary response: Having always the same form: not varying.

Q 2-22 It is somewhat obsolete to use a building trap nowadays.

The word **obsolete,** as used in this sentence, means:

(A) out-of-date
(B) repaired

(C) reconditioned
(D) worn out

Universal Source: Dictionary

Answer: A
Dictionary response: No longer in use, or outmoded.

Q 2-23 Expansion joints are seldom used in the water piping of a single-family residence.

The word **seldom,** as used in this sentence, means:

(A) never
(B) always

(C) rarely
(D) often

Universal Source: Dictionary

Answer: C
Dictionary response: Not often, rarely.

Q 2-24 Of the following plumbing terms, the one most correctly defined, according to the Code, is:

(A) "soil pipe" shall mean any pipe manufactured of PVC schedule 40 which conveys waste to the building drain
(B) "sub-drain" shall mean that portion of a drainage system installed below the first floor level

(C) "indirect waste pipe" shall mean a waste pipe which fails to connect directly to the building drainage system
(D) "stack" shall mean any line of soil, waste or vent piping

---

Standard Plumbing Code	Uniform Plumbing Code

Answer: C
Code response: An indirect waste pipe is a pipe that does not connect directly with the drainage system.

Answer: C
Code response: An indirect waste pipe is a pipe that does not connect directly with the drainage system.

Q 2-25 Of the following plumbing terms, the one most correctly defined, according to the Code, is:

(A) "developed length" shall mean its length along the centerline of the pipe and fittings
(B) "continuous waste" shall mean a drain connecting several similar fixtures to one drain pipe

(C) "leader" shall mean any pipe conveying storm water to an approved disposal area
(D) "branch" shall mean any part of the piping system including fittings and risers

---

Standard Plumbing Code	Uniform Plumbing Code

Answer: A
Code response: The developed length of a pipe is its length along the centerline of the pipe and fittings.

Answer: A
Code response: The developed length of a pipe is its length along the centerline of the pipe and fittings.

Q **2-26** Which of the following plumbing term definitions is most nearly correct?

(A) a "septic tank" is a watertight receptacle designed so as to receive solids and allow the liquids to discharge into the soil

(B) "special wastes" are wastes which require the use of indirect waste piping

(C) a "sump" is an approved tank which is located above the normal grade of the gravity system of a building and must be emptied by mechanical means

(D) a "seepage pit" is a pit with a perforated lining into which the effluent from a septic tank is discharged

Standard Plumbing Code

Answer: A
Code response: A septic tank is a watertight receptacle which receives the discharge of a drainage system, designed so as to separate solids and allow the liquids to discharge into the soil outside the tank through a system of open-joint piping.

Uniform Plumbing Code

Answer: A
Code response: A septic tank is a watertight receptacle which receives the discharge of a drainage system, designed so as to retain solids and allow the liquids to discharge into the soil outside the tank through a system of open-joint piping.

Q **2-27** A device which prevents water of questionable origin from entering the distributing pipes of a potable supply of water is defined by Code as:

(A) check valve
(B) cross-connection

(C) backflow preventer
(D) air gap

Standard Plumbing Code

Answer: C
Code response: A backflow device prevents the reverse flow of water into the potable water supply system.

Uniform Plumbing Code

Answer: C
Code response: A backflow preventer is a device to prevent backflow into the potable water supply system.

Q **2-28** The Code defines a sewer which receives storm water, sewage and liquid waste as a:

(A) private sewer
(B) combined sewer

(C) public sewer
(D) special sewer

Standard Plumbing Code

Answer: B
Code response: A combined building sewer receives storm water, sewage and liquid waste.

Uniform Plumbing Code

Answer: B
Code response: Roofs, inner courts, or similar areas having rain water drains, shall not be connected to the drainage system.

Note: The definition, as given by the Uniform Plumbing Code, is considered a combined sewer. Most Codes prohibit the installation of a combined sewer unless first approved by the Administrative Authority.

2-29 The Code defines _____ as any approved on-site sewage disposal system used in lieu of a standard subsurface system.

(A) aquifer
(B) pit privy

(C) chemical toilet
(D) alternate system

Standard Plumbing Code

Uniform Plumbing Code

Answer: D
Code response: The Administrative Authority. may approve an alternate system.

Answer: D
Code response: Alternate systems may be used only by special permission of the Administrative Authority.

2-30 Back-siphonage is the reverse flow of questionable water from a plumbing fixture into a potable water supply pipe due to _____ in such pipe.

(A) negative pressure
(B) positive pressure

(C) true siphon action
(D) capillary attraction

Standard Plumbing Code

Uniform Plumbing Code

Answer: A
Code response: Back-siphonage is the flowing back of water from a plumbing fixture into a water supply pipe due to a negative pressure in such pipe.

Answer: A
Code response: Back-siphonage is the flowing back of used water from a plumbing fixture into a water supply pipe due to a negative pressure in such pipe.

2-31 The Code defines a closed water piping system as a system of piping where a _____ prevents the free return of water to the public water main.

(A) gate valve
(B) ball valve

(C) butterfly valve
(D) check valve

Standard Plumbing Code

Uniform Plumbing Code

Answer: D
Code response: A check valve is used in a closed water piping system to prevent the free return of water to the water main.

Answer: D
Code response: Where backflow can occur, an approved backflow prevention device shall be installed in the supply line.

2-32 Siphonage of the contents in a hot water storage tank may occur when the main water supply pipe to a building is:

(A) equipped with a pressure regulator

(B) oversized to overcome an excessive length or run

(C) closed, and a hose bib on the supply line located at a lower elevation is opened
(D) under excessive pressure from the public water main

Standard Plumbing Code

Uniform Plumbing Code

Answer: C
Code response: Back-siphonage is the flowing back of water from a plumbing fixture (water heater) into a water supply pipe.

Answer: C
Code response: Back-siphonage is the flowing back of water from a plumbing fixture (water heater) into a water supply pipe.

Q 2-33 According to Code, the distance "X" across the 4-inch pipe shown in Figure 2-3 is defined as the:

Figure 2-3

(A) circumference
(B) diameter

(C) cross-section
(D) area

Standard Plumbing Code

Uniform Plumbing Code

Answer: B
Code response: The term "diameter" is the nominal diameter as designated commercially.

Answer: B
Code response: The term "diameter" is the nominal diameter as designated commercially.

Note: Neither Code specifically defines "diameter." The universal source, a dictionary, defines the term "diameter" as a straight line that passes through the center of a circle and divides it in half.

Q 2-34 In Figure 2-4, the Code defines pipe "A" as a:

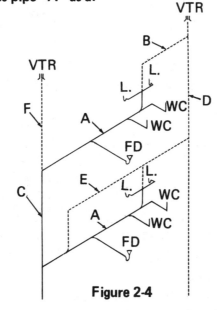

Figure 2-4

(A) waste pipe
(B) fixture branch

(C) continuous waste
(D) horizontal branch

Standard Plumbing Code

Uniform Plumbing Code

Answer: D
Code response: A horizontal branch is a drain pipe extending laterally from a soil or waste stack, which receives the discharge from one or more fixture drains and conducts it to the soil or waste stack.

Answer: D
Code response: A horizontal branch is a drain pipe extending laterally from a soil or waste stack, which receives the discharge from one or more fixture drains and conducts it to the soil or waste stack.

Q 2-35 In Figure 2-4, the Code defines pipe "B" as a:

(A) combined vent
(B) circuit vent

(C) relief vent
(D) common vent

Standard Plumbing Code

Uniform Plumbing Code

Answer: B
Code response: A circuit vent is a branch vent that serves two or more traps and extends from in front of the last fixture connection of a horizontal branch to the vent stack.

Answer: B
Code response: A circuit vent is a branch vent that serves two or more traps and extends from in front of the last fixture connection of a horizontal branch to the vent stack.

Q 2-36 In Figure 2-4, the Code defines pipe "C" as a:

(A) soil pipe
(B) waste pipe

(C) soil stack
(D) waste stack

Standard Plumbing Code

Uniform Plumbing Code

Answer: C
Code response: A stack is the vertical main of a system of soil, waste, or vent piping.

Answer: C
Code response: A stack is the vertical main of a system of soil, waste, or vent piping extending one or more stories in height.

Q 2-37 The Code defines pipe "D" in Figure 2-4 as a:

(A) vent stack
(B) stack vent

(C) revent pipe
(D) dry vent

Standard Plumbing Code

Uniform Plumbing Code

Answer: A
Code response: A vent stack is a vertical vent pipe installed to provide circulation of air to and from a drainage system.

Answer: A
Code response: A vent stack is a vertical vent pipe installed to provide circulation of air to and from a discharge system.

Q 2-38 The Code defines pipe "E" in Figure 2-4 as a:

(A) vertical to a horizontal vent pipe
(B) relief or revent pipe

(C) main or secondary vent pipe
(D) individual or group vent pipe

Standard Plumbing Code

Uniform Plumbing Code

Answer: B
Code response: A revent pipe is that part of a vent pipe line which connects directly with individual or a group of wastes and extends to the main or branch vent pipe.

Answer: B
Code reponse: A relief vent is a vent, the primary function of which is to act as an auxiliary vent on a specially designed system.

Q 2-39 The Code defines pipe "F" in Figure 2-4 as:

(A) a vent stack
(B) an individual vent

(C) a continuous vent
(D) a stack vent

Standard Plumbing Code

Uniform Plumbing Code

Answer: D
Code response: A stack vent is the extension of a soil or waste stack above the highest horizontal drain connected to the stack.

Answer: D
Code response: A stack vent is the extension of a soil or waste stack above the highest horizontal drain connected to the stack.

Q 2-40 Figure 2-5 shows a special venting method that is defined by Code as a:

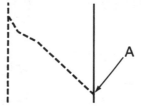

Figure 2-5

(A) yoke vent
(B) relief vent

(C) unit vent
(D) revent pipe

Standard Plumbing Code

Uniform Plumbing Code

Answer: A
Code response: A yoke vent is a pipe connecting upward from a soil or waste stack to a vent stack.

Answer: A
Code response: A yoke vent is a pipe connecting upward from a soil or waste stack to a vent stack.

Q 2-41 Pipe "A" in Figure 2-6, is defined by Code as a:

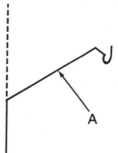

Figure 2-6

(A) waste pipe
(B) fixture arm

(C) fixture branch
(D) fixture drain

Standard Plumbing Code

Uniform Plumbing Code

Answer: D
Code response: A fixture drain is the drain from the fixture trap to the junction with any other drain pipe.

Answer: D
Code response: A fixture drain is the drain from the fixture trap to the junction with any other drain pipe.

Q. 2-42 The Code defines a pipe which makes an angle of not more than 45 degrees with the vertical as:

(A) straight
(B) vertical

(C) unacceptable
(D) graded

Standard Plumbing Code

Uniform Plumbing Code

Answer: B
Code response: A vertical pipe is any pipe or fitting which makes an angle of not more than 45 degrees with the vertical.

Answer: B
Code response: A vertical pipe is any pipe or fitting which makes an angle of not more than 45 degrees with the vertical.

Q. 2-43 The Code defines a trap as a fitting or device designed to:

(A) serve as a passage for waste water
(B) retain a liquid seal

(C) prevent sewer gases from entering the building
(D) retain and prevent certain harmful substances from entering the building drainage system

Standard Plumbing Code

Uniform Plumbing Code

Answer: C
Code response: A trap is a fitting or device so designed as to provide a liquid seal which will prevent the back-passage of air.

Answer: C
Code response: A trap is a fitting or device so designed as to provide a liquid seal which will prevent the back-passage of air.

Q. 2-44 The Code defines the depth of a trap seal as being measured from the:

(A) crown weir to the top of the dip of a trap
(B) crown weir to the bottom of the dip of a trap

(C) fixture tailpiece to the crown weir of a trap
(D) fixture tailpiece to the top of the dip of a trap

Standard Plumbing Code

Uniform Plumbing Code

Answer: A
Code response: The trap seal is the maximum vertical depth of liquid, measured between the crown weir and the top of the dip of the trap.

Answer: A
Code response: The trap seal is the maximum vertical depth of liquid, measured between the crown weir and the top of the dip of the trap.

Q 2-45 Figure 2-7 shows a method that can be used for venting several fixtures. The Code defines it as a:

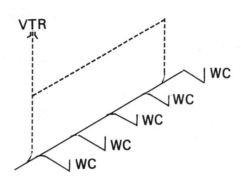

Figure 2-7

(A) revent
(B) unit vent

(C) local vent
(D) loop vent

Standard Plumbing Code	Uniform Plumbing Code
Answer: D Code response: A loop vent loops back and connects with the stack vent.	**Answer: D** Code response: A loop vent loops back and connects with the stack vent.

Q 2-46 The term "fixture unit," as defined by Code, is based upon the rate of discharge from a:

(A) water cooler
(B) bar sink

(C) bathtub
(D) lavatory

Standard Plumbing Code	Uniform Plumbing Code
Answer: D Code response: A fixture unit (flow rate) is a quantity in terms of which the load-producing effects on a plumbing system of different kinds of fixtures are expressed on some arbitrarily chosen scale.	**Answer: D** Code response: A fixture unit (flow rate) is a quantity in terms of which the load-producing effects on a plumbing system of different kinds of fixtures are expressed on some arbitrarily chosen scale.

Note: The nationally accepted standard for determining a fixture unit is based on the flow rate of a lavatory. Most lavatories are rated as 1 fixture unit, which is equal to 7½ gallons flow per minute.

Q 2-47 The primary purpose of a vent system, as defined by Code, is a pipe or pipes installed to:

(A) prevent capillary attraction from occurring

(B) prevent negative pressure from developing within the drainage system

(C) prevent positive pressure from being created within the drainage system

(D) provide a circulation of air within the drainage system

Standard Plumbing Code	Uniform Plumbing Code
Answer: D Code response: A vent system is a pipe or pipes installed to provide a circulation of air within such system.	**Answer: D** Code response: A vent system is a pipe or pipes installed to provide a circulation of air within such system.

Q 2-48 A ____ does **not** constitute part of a water supply system as defined by Code.

(A) hydropneumatic tank

(B) pipe strap

(C) water softener

(D) ½-inch by ¾-inch reducing coupling

Standard Plumbing Code

Uniform Plumbing Code

Answer: B

Code response: The water supply system consists of the water supply piping, connecting pipes, fittings, control valves and all appurtenances connected thereto.

Answer: B

Code response: The water supply of a building consists of the supply pipe, the water distributing pipes, fittings, control valves and all appurtenances carrying or supplying potable water to the building.

Q 2-49 A Durham System, as defined by Code, is a system where:

(A) piping and fittings are welded

(B) piping and fittings are of the flare type

(C) piping is threaded and fittings are of the recessed type

(D) piping and fittings are of the flanged type

Standard Plumbing Code

Uniform Plumbing Code

Answer: C

Code response: Durham System is a term used to describe soil or waste systems where all piping is threaded, using recessed drainage fittings.

Answer: C

Code response: Durham System is a term used to describe soil or waste systems where all piping is threaded, using recessed drainage fittings.

Q 2-50 A drainage system, as defined by Code, includes all piping within public or private premises which conveys:

(A) sewage

(B) rain water

(C) liquid wastes

(D) all of the above

Standard Plumbing Code

Uniform Plumbing Code

Answer: D

Code response: A drainage system conveys sewage, rain water and other liquid wastes.

Answer: D

Code response: A drainage system conveys sewage or other liquid wastes.

Note: The Uniform Plumbing Code does not specifically address rain water as being a liquid conveyed by a drainage system. Rain water is considered a liquid waste and is conveyed to a point of disposal by the building drainage system (although, in most instances, there is a separate system for rain water). Answer "D" is correct.

Q 2-51 The gradient of a horizontal line of pipe, as defined by Code, is usually expressed in terms of:

(A) a fraction of an inch

(B) a fraction of a foot

(C) being below the frost line

(D) its extreme depth

Standard Plumbing Code

Uniform Plumbing Code

Answer: A

Code response: Grade is the slope or fall of a line of pipe, usually expressed in a fraction of an inch per foot.

Answer: A

Code response: Grade is the slope or fall of a line of pipe, usually expressed in a fraction of an inch or percentage slope per foot.

Q 2-52 Suppose the date of the Code you are using has just become effective today. Let's say you are adding a bathroom to a residence. The Code defines the existing plumbing system as:

(A) work installed by another plumber
(B) work installed prior to the effective date of this Code

(C) a private drainage system
(D) work installed after the effective date of this Code

Standard Plumbing Code	Uniform Plumbing Code

Answer: B
Code response: Existing work is plumbing installed prior to the effective date of this Code.

Answer: B
Code response: Existing work is plumbing installed prior to the effective date of this Code.

Q 2-53 The Code defines any horizontal pipe or fittings as being installed:

(A) with a pitch of not more than 1/8-inch fall per foot
(B) with a slope in line of a pipe, in reference to a horizontal plane

(C) having an angle of more than 45 degrees with the vertical
(D) having an angle of less than 45 degrees with the vertical

---

Standard Plumbing Code	Uniform Plumbing Code

Answer: C
Code response: Horizontal pipe is defined as any pipe or fitting making an angle of more than 45 degrees with the vertical.

Answer: C
Code response: Horizontal pipe is defined as any pipe or fitting installed in a horizontal position making an angle of not more than 45 degrees with the horizontal.

Note: Each Code explains it from a different reference. The Standard Plumbing Code uses the **vertical** as its point of reference; the Uniform Plumbing Code uses the **horizontal** as its point of reference. However, you can see they both have the same meaning. The correct response is "C."

Q 2-54 Of the following piping installations, the one that is not defined by Code as a "stack" when installed as a vertical main is:

(A) soil piping
(B) water piping

(C) waste piping
(D) vent piping

Standard Plumbing Code	Uniform Plumbing Code

Answer: B
Code response: The vertical main of a system of soil, waste, or vent piping is defined as a stack.

Answer: B
Code response: A stack is the vertical main of a system of soil, waste or vent piping extending through one or more stories.

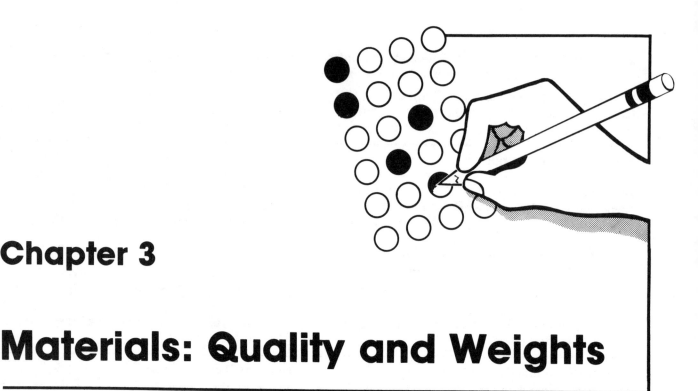

Chapter 3

Materials: Quality and Weights

The plumbing code protects the public health, welfare, and safety by requiring proper design, installation and maintenance of plumbing systems. Good workmanship and design are useless, however, if the materials used don't provide satisfactory service. That's why the code sets minimum standards for quality and weight of plumbing materials.

The code lists minimum standards for all materials, fixtures or devices used in the construction of plumbing and drainage systems. Materials must be free from defects. All pipe, pipe fittings and fixtures must be listed or labeled by a listing agency. The marking of materials is actually done by the manufacturer.

New materials and manufacturing methods have resulted in new products that are accepted as substitutes for older code-approved materials. The administrative authority (government agency that adopted and administers the code) may approve alternate materials so long as they meet these requirements:

•   The proposed design must be satisfactory and comply with the intent of the code.

•   The material offered must be suitable for the purpose intended.

•   The material must be at least equivalent to that prescribed by the code in quality, strength, effectiveness, durability and safety.

•   The proposed methods of installation must conform to other acceptable nationally-recognized plumbing standards.

State and local exams include many questions from this section of the code. You'll need a good working knowledge of material quality and weights to pass the exam and to do your work properly as a licensed plumber.

Q 3-1 The Code requires that sheet lead used in the construction of safe pans for shower compartments weigh no less than _____ pounds per square foot.

(A) 2
(B) 2½

(C) 3
(D) 4

Standard Plumbing Code

Uniform Plumbing Code

Answer: D
Code response: Sheet lead for safe pans shall not be less than 4 pounds per square foot.

Answer: D
Code response: Sheet lead shall weigh no less than 4 pounds per square foot for safe pans.

Q 3-2 The Code requires that soldering bushings used with 2½-inch pipe have a minimum weight of _____ ounces.

(A) 6
(B) 8

(C) 14
(D) 22

Standard Plumbing Code

Uniform Plumbing Code

Answer: D
Code response: Solder bushings used with 2½-inch pipe shall weigh no less than 1 lb. 6 oz. each.

Answer: D
Code response: Solder bushings used with 2½-inch inch pipe shall have a minimum weight of 1 lb. 6 oz.

Q 3-3 Code specifies that soldering bushings, when permitted, are usually manufactured of:

(A) zeolite
(B) brass

(C) zinc
(D) antimony

Standard Plumbing Code

Uniform Plumbing Code

Answer: B
Code response: Soldering bushings, where permitted, shall be of red brass.

Answer: B
Code response: Soldering bushings shall be of bronze or copper.

Note: Copper has various alloys, such as brass and bronze, all of which are acceptable for solder bushings. Answer "B" is correct.

Q 3-4 According to Code, seamless copper water tubing used for water distribution must meet the _____ standards.

(A) ASTM
(B) ANSI

(C) PS
(D) AWWA

Standard Plumbing Code

Uniform Plumbing Code

Answer: A
Code response: The **American Society for Testing and Materials (ASTM)** is the adopted standard used by this Code for seamless copper water tube.

Answer: A
Code response: The **American Society for Testing and Materials (ASTM)** is the adopted standard used by this Code for copper plumbing tube and fittings.

Note: See "Materials for Plumbing Installation" or "Material Standards," as listed in your Code book.

Q 3-5 According to Code, the material used in the manufacture of cast brass flared fittings is classified as being:

(A) ferrous

(B) nonmetallic

(C) nonferrous

(D) metallic

Standard Plumbing Code

Uniform Plumbing Code

Answer: C

Code response: Cast bronze fittings for flared copper tube is listed by the Code as being non-ferrous.

Answer: C

Code response: Cast bronze fittings for flared copper tubes is listed by the Code as being non-ferrous.

Note: See "Materials for Plumbing Installation" or "Material Standards," as listed in your Code book.

Q 3-6 Code demands that caulking ferrules be manufactured from:

(A) copper

(B) cast iron

(C) black steel

(D) wrought iron

Standard Plumbing Code

Uniform Plumbing Code

Answer: A

Code response: Caulking ferrules shall be manufactured of bronze or copper.

Answer: A

Code response: Caulking ferrules shall be manufactured of bronze or copper.

Q 3-7 Caulked-on floor flanges for water closets, according to Code, must have a minimum caulking depth of _____ .

(A) 1 inch

(B) 1½ inches

(C) 2 inches

(D) 2½ inches

Standard Plumbing Code

Uniform Plumbing Code

Answer: C

Code response: Floor flanges shall have not less than 2-inch caulked depth.

Answer: C

Code response: Caulked-on flanges shall be not less than 2 inches in overall depth.

Q 3-8 To meet Code, floor flanges for water closets manufactured of cast iron must have a minimum thickness of:

(A) 1/16-inch

(B) 1/8- inch

(C) 3/16-inch

(D) 1/4- inch

Standard Plumbing Code

Uniform Plumbing Code

Answer: D

Code response: Cast-iron floor flanges for water closets shall be not less than ¼-inch-thick.

Answer: D

Code response: Caulked-on flanges shall be not less than ¼-inch-thick.

Q **3-9** A 3-inch bronze caulking ferrule must have a minimum length of _____, according to the Code.

(A) 3½ inches
(B) 4   inches

(C) 4½ inches
(D) 5   inches

Standard Plumbing Code

Uniform Plumbing Code

Answer: C
Code response: Caulking ferrules shall be 4½ inches in length.

Answer: C
Code response: Caulking ferrules shall be 4½ inches in length.

Q **3-10** Where sheet lead is used to make watertight vent terminals, the Code requires that such sheet lead weigh no less than _____ pounds per square foot.

(A) 2
(B) 2½

(C) 3
(D) 3½

Standard Plumbing Code

Uniform Plumbing Code

Answer: C
Code response: Sheet lead for flashing of vent terminals - not less than 3 pounds per square foot.

Answer: C
Code response: Sheet lead for flashing vent terminals - not less than 3 pounds per square foot.

Q **3-11** Of the following standards, the one that is <u>not</u> a Code-adopted standard for cast-iron pipe and fittings is:

(A) ANSI
(B) ASME

(C) ASTM
(D) FS

Standard Plumbing Code

Uniform Plumbing Code

Answer: B
Code response: **American Society of Mechanical Engineers** (ASME) is not listed as an adopted standard for cast-iron pipe fittings.

Answer: B
Code response: **American Society of Mechanical Engineers** (ASME) is not listed as an adopted standard for cast-iron pipe and fittings.

Q **3-12** The Code states that in a cast-iron drainage system where lead closet bends are used, the water-closet flange must be secured to the water-closet stub by:

(A) oakum and poured lead
(B) soldering

(C) lead wool
(D) a graphite gasket

Standard Plumbing Code

Uniform Plumbing Code

Answer: B
Code response: Closet flanges shall be soldered to lead bends.

Answer: B
Code response: Closet flanges shall be burned or soldered to lead bends or stubs.

3-13 The Administrative Authority may approve alternate materials that are not specifically defined in the Code, providing it is clear that:

(A) the Code is not applicable

(B) minimum standards are equal

(C) the manufacturer's instructions for installation are adhered to

(D) the manufacturer will assume full responsibility for his material

Standard Plumbing Code

Uniform Plumbing Code

Answer: B
Code response: The Plumbing Official shall approve alternate materials if he finds they are equivalent to prescribed standards.

Answer: B
Code response: The Administrative Authority may approve alternate materials after he is satisfied as to their adequacy.

3-14 Polybutylene piping and fittings, according to Code, would most likely be used in:

(A) a water distribution system
(B) special wastes systems

(C) indirect waste piping
(D) wet vented systems

Standard Plumbing Code

Uniform Plumbing Code

Answer: A
Code response: Materials for water distribution pipes may be polybutylene.

Answer: A
Code response: Polybutylene water piping and fittings may be used for water distribution systems.

3-15 Acrylonitrile-butadiene-styrene piping and fittings, according to Code, would most generally be used for:

(A) above-ground water piping
(B) swimming pool pressure piping

(C) drainage systems
(D) natural gas yard piping

Standard Plumbing Code

Uniform Plumbing Code

Answer: C
Code response: The standards for ABS piping and materials are generally limited to and used most often for the building drainage system.

Answer: C
Code response: ABS piping is approved for use in drainage systems only.

3-16 According to Code, DWV copper drainage tubing must meet the standards of:

(A) the American National Standards Institute

(B) the American Water Works Association

(C) the American Society for Testing and Materials

(D) Federal Specifications, as published by the Federal Specifications Board

Standard Plumbing Code

Uniform Plumbing Code

Answer: C
Code response: The **American Society for Testing and Materials** is the adopted standard used by this Code for copper drainage tubing (DWV).

Answer: C
Code response: The **American Society for Testing and Materials** is the adopted standard used by this Code for copper drainage tubing (DWV).

Note: See "Materials for Plumbing Installation" or "Material Standards," as listed in your Code book.

Q 3-17 **The material used in the manufacture of galvanized steel piping, according to Code, is classified as being:**

(A) ferrous

(B) nonferrous

(C) carbon

(D) silicon and manganese

Standard Plumbing Code

Uniform Plumbing Code

Answer: A
Code response: Galvanized steel piping is listed by the Code as being ferrous.

Answer: A
Code response: Galvanized steel piping is listed by the Code as being ferrous.

Note: See "Materials for Plumbing Installation" or "Material Standards," as listed in your Code book.

Q 3-18 **The Code makes it clear that all materials used in the installation of a plumbing and drainage system must conform to:**

(A) manufacturer's standards

(B) all alternate materials available

(C) approved maximum standards

(D) approved minimum standards

Standard Plumbing Code

Uniform Plumbing Code

Answer: D
Code response: The materials listed in this chapter shall conform at least to the standards cited.

Answer: D
Code response: Standards listed or referred to in this chapter cover materials which will conform to the requirements of this Code.

Note: Standards listed, referred to, or cited are the minimum requirements set by Code.

Q 3-19 **Of the following materials, the one approved by Code to secure water closets to tile floor is:**

(A) brass

(B) bismuth

(C) babbitt metal

(D) tungsten

Standard Plumbing Code

Uniform Plumbing Code

Answer: A
Code response: Screws and bolts for water closets shall be of brass.

Answer: A
Code response: Closet screws, bolts, washers shall be of brass.

Q 3-20 **Of the following materials, the one approved by Code for cleanout plugs in a cast-iron drainage system is:**

(A) wrought iron

(B) cast steel

(C) brass

(D) blister steel

Standard Plumbing Code

Uniform Plumbing Code

Answer: C
Code response: Cleanouts shall have plugs of brass.

Answer: C
Code response: Each cleanout fitting for cast-iron pipe shall consist of a brass plug.

3-21 **The wall thickness of lead bends used in a plumbing system, according to Code, must be _____ inch thick.**

(A) 1/16
(B) 1/8

(C) 3/16
(D) 1/4

Standard Plumbing Code

Uniform Plumbing Code

Answer: B
Code response: Lead bends shall not be less than 1/8-inch wall thickness.

Answer: B
Code response: Lead bends shall not be less than 1/8-inch wall thickness.

3-22 **The Code requirement is that 4-inch pipe size caulking ferrules weigh a minimum of _____ each.**

(A) 1 lb. 6 oz.
(B) 1 lb. 12 oz.

(C) 2 lb. 3 oz.
(D) 2 lb. 8 oz.

Standard Plumbing Code

Uniform Plumbing Code

Answer: D
Code response: Minimum weight each for 4-inch pipe size caulking ferrules is 2 lb. 8 oz.

Answer: D
Code response: Minimum weight each for 4-inch pipe size caulking ferrules is 2 lb. 8 oz.

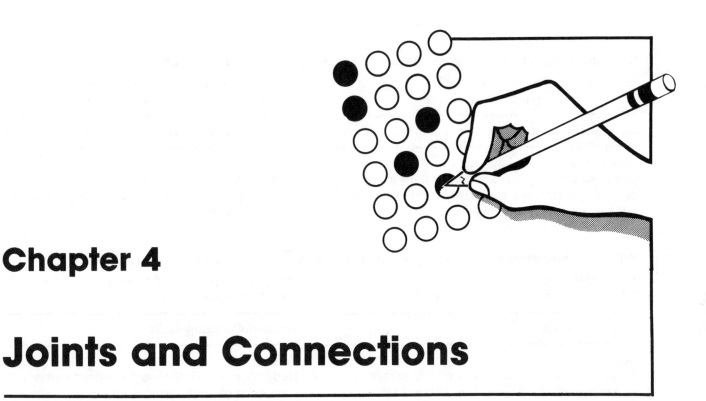

Chapter 4

Joints and Connections

Every plumbing system requires the installation of many fittings. Fittings, of course, require joints and connections. The code requires that joints and connections be gas-tight and watertight during the pressure test and when used as intended.

The type of joint depends on the materials used. Joints regulated by code include: caulked joints, threaded joints, wiped joints, soldered joints, flared joints, cement mortar joints, burned lead joints, asbestos cement sewer pipe joints, bituminized fiber pipe joints, solvent cement plastic pipe joints, mechanical joints, flexible compression factory-fabricated joints, molded rubber coupling joints, elastomeric gasketed and rubber-ring joints, and joints made by brazing and welding.

The code regulates the use of all common and special joints, flanged fixture connections and the waterproofing of openings. The code also prohibits the use of certain types of joints and connections.

It's important for you to be well acquainted with joints and connections acceptable for use in a plumbing system. The questions in this chapter test your knowledge of this section of the code. If you miss any of these questions, review your local code or the appropriate reference books.

Q **4-1 Code mandates that joints and connections in the plumbing system be gastight and watertight for:**

(A)   the intended use
(B)   the working pressure available

(C)   the pressures required by test
(D)   any fluctuation of pressure

Standard Plumbing Code

Answer: C
Code response: Joints and connections in the plumbing system shall be gastight and watertight for the pressures required by test.

Uniform Plumbing Code

Answer: C
Code response: Joints and connections in the plumbing system shall be gastight and watertight for the pressures required by test.

Q **4-2 When cement mortar joints are permitted by Code, the mortar mix must be composed of:**

(A)   2 parts cement and 1 part sand
(B)   2 parts cement and 2 parts sand

(C)   1 part cement and 1 part sand
(D)   1 part cement and 2 parts sand

Standard Plumbing Code

Answer: D
Code response: A thoroughly mixed mortar composed of 1 part cement and 2 parts sand.

Uniform Plumbing Code

Answer: D
Code response: A thoroughly mixed mortar composed of 1 part cement and 2 parts sand.

Q **4-3 To meet the Code, 1-inch galvanized steel pipe threads must be:**

(A)   wiped clean of cutting oil
(B)   standard taper threads

(C)   checked for burrs
(D)   a minimum of 1-inch long

Standard Plumbing Code

Answer: B
Code response: Threaded iron pipe joints shall conform to American National Taper pipe thread.

Uniform Plumbing Code

Answer: B
Code response: Threads on iron pipe size (I.P.S.) pipe shall be standard taper threads.

Q **4-4 According to Code, _____ joints may be used where necessary to provide for expansion and contraction of the pipes and must be accessible.**

(A)   slip
(B)   flanged

(C)   brazed
(D)   swing

Standard Plumbing Code

Answer: D
Code response: Expansion (swing) joints must be accessible and may be used to provide for expansion and contraction of the pipes.

Uniform Plumbing Code

Answer: D
Code response: Expansion (swing) joints shall be accessible and may be used where necessary to provide for expansion and contraction of the pipes.

4-5 The Code will permit the installation of elastomeric compression type joints as an alternative to _____ on hub and spigot cast-iron pipe.

(A) lead and oakum joints
(B) stainless steel no-hub couplings

(C) cement and oakum joints
(D) hot poured compound joints

Standard Plumbing Code

Uniform Plumbing Code

Answer: A
Code response: Elastomeric compression type joints may be used for joining hub and spigot cast-iron soil pipe as an alternative to lead and oakum joints.

Answer: A
Code response: Elastomeric compression type gasket joints may be used for joining hub and spigot cast-iron pipe as an alternative to lead and oakum joints.

4-6 In water piping, the Code permits the use of slip joints:

(A) if they are accessible
(B) if they are used on a lavatory supply

(C) only on hot water lines for expansion purposes
(D) where piping materials are dissimilar

Standard Plumbing Code

Uniform Plumbing Code

Answer: B
Code response: Slip joints in water piping may only be used on the exposed fixture supply.

Answer: B
Code response: In water piping, slip joints may be used only on the exposed fixture supply.

4-7 In making a vertical caulked joint on a 4-inch cast-iron soil pipe, oakum must first be firmly packed into the joint. The depth of lead, according to Code, must be at least _____ inches.

(A) ½
(B) ¾

(C) 1
(D) 1¼

Standard Plumbing Code

Uniform Plumbing Code

Answer: C
Code response: Caulked joints for cast-iron hub-and-spigot soil pipe shall be filled with molten lead not less than 1 inch deep.

Answer: C
Code response: Caulked joints for cast-iron bell-and-spigot soil pipe shall be filled with molten lead to a depth of not less than 1 inch.

4-8 A 1-inch horizontal hot water copper pipe which exceeds 100 feet in length must compensate for its expansion by being:

(A) well insulated to meet energy requirements

(B) supported with straps not less than 8 feet apart

(C) graded at not less than 1/16-inch per foot toward boiler
(D) protected with swing joints

Standard Plumbing Code

Uniform Plumbing Code

Answer: D
Code response: Provisions shall be made for expansion of hot water piping.

Answer: D
Code response: Approved provisions shall be made for expansion of hot water piping.

Q 4-9 Of the following materials, the one that is used to make up a threaded joint between copper tubing and galvanized steel piping is:

(A) brass male adapter

(B) dresser coupling

(C) approved kafer fitting

(D) galvanized extension piece

Standard Plumbing Code

Uniform Plumbing Code

Answer: A

Code response: Joints from copper tubing to threaded pipe shall be made by the use of bronze adapter fittings.

Answer: A

Code response: Joints from copper tubing to threaded pipe shall be made by the use of brass adapter fittings.

Q 4-10 Of the following joints, the one that is Code-approved for lead pipe is the:

(A) sisson joint

(B) burned joint

(C) wiped joint

(D) soldered joint

Standard Plumbing Code

Uniform Plumbing Code

Answer: C

Code response: Joints between lead pipe and other type metals shall be wiped joints.

Answer: C

Code response: Joints between lead pipe and other type approved metals shall be wiped joints.

Q 4-11 Caulking ferrules, according to Code, are approved as a connection between:

(A) K copper pipe and black steel pipe

(B) cast-iron pipe and brass pipe

(C) cast-iron pipe and lead pipe

(D) cast-iron pipe and wrought-iron pipe

Standard Plumbing Code

Uniform Plumbing Code

Answer: C

Code response: Joints between lead and cast-iron pipe shall be made by means of wiped joints and a caulking ferrule.

Answer: C

Code response: Joints between lead and cast-iron pipe shall be made by means of wiped joints and a caulking ferrule.

Q 4-12 An architect specifies a cast-iron drainage system with 4-inch lead closet stubs. A full wiped joint is made between the 4-inch lead pipe and the 4-inch brass ferrule. The minimum dimensions of each wiped joint, according to Code, must be:

(A) 1-1/8 inches long by 4 inches in diameter

(B) 1¼ inches long by 4¼ inches in diameter

(C) 1½ inches long by 4½ inches in diameter

(D) 1½ inches long by 5½ inches in diameter

Standard Plumbing Code

Uniform Plumbing Code

Answer: C

Code response: Lead bends shall not be less than 1/8-inch wall thickness. Wiped joints shall have an exposed surface on each side of a joint not less than ¾-inch and at least as thick as the material being joined.

Answer: C

Code response: Lead bends shall not be less than 1/8-inch wall thickness. Wiped joints shall have an exposed surface on each side of a joint not less than ¾-inch and at least as thick as the material being joined.

Q 4-13 To connect two pieces of galvanized steel pipe in the water supply system, the most acceptable Code-approved fitting is a:

(A) black steel dresser coupling
(B) brass compression coupling

(C) galvanized union with graphite gasket
(D) galvanized metal-to-metal ground seat union

Standard Plumbing Code

Uniform Plumbing Code

Answer: D
Code response: Metal-to-metal ground seat unions shall be used in water distribution systems.

Answer: D
Code response: Approved unions may be used at any point in the water supply system.

Q 4-14 According to Code, threaded plastic pipe must not be less than schedule ____ minimum wall thickness.

(A) 40
(B) 60

(C) 80
(D) 120

Standard Plumbing Code

Uniform Plumbing Code

Answer: ____
Code response: Joints for plastic pipe shall be the insert type, solvent-cemented or may be hot- or cold-flared, as recommended by the manufacturer.

Answer: C
Code response: Threaded plastic pipe shall be schedule 80 minimum wall thickness.

Note: The Standard Plumbing Code does not include threaded plastic pipe as an approved connection. Check local Code to see if it addresses this particular question.

Q 4-15 To make a bell-and-spigot joint between cast-iron pipe and vitrified clay pipe, the only Code-approved method to use is:

(A) molded rubber coupling joints
(B) oakum and poured lead

(C) lead wool
(D) a rich mortar mix

Standard Plumbing Code

Uniform Plumbing Code

Answer: A
Code response: Flexible couplings may be used to adapt any two of the following pipes of the same O.D.-vitrified clay pipe and cast-iron pipe, for example.

Answer: A
Code response: Joints between vitrified clay pipe and metal pipe may be made with molded rubber coupling joints.

Q 4-16 Horizontal swing joints or pipe loops on long runs of hot water piping are generally installed (to satisfy Code requirements) for the purpose of:

(A) relieving the possibility of water hammer in the piping
(B) relieving the density of the water being conveyed by the pipe

(C) allowing for movement of pipe due to temperature changes
(D) avoiding viscosity buildup in the piping

Standard Plumbing Code

Uniform Plumbing Code

Answer: C
Code response: Expansion joints may be used where necessary to provide for expansion of the pipes.

Answer: C
Code response: Provisions shall be made for expansion in hot water piping.

4-17 Recessed threaded drainage ells are used to run a horizontal offset in a vent branch. The fittings are properly tapped and recessed to allow a grade of ¼-inch per foot. The developed length of this offset is 14 feet. The difference in elevation from end to end of this offset should be _____ inches.

(A) 1½   (C) 2½

(B) 2   (D) 3½

Standard Plumbing Code	Uniform Plumbing Code
Answer: D	**Answer: D**
Code response: Fittings on the drainage system shall be of the recessed drainage type.	Code response: Drainage fittings shall be constructed of the recessed type and tapped so as to allow ¼-inch per foot grade.

Note: Threaded drainage fittings are generally tapped with a pitch by the manufacturer to permit approximately ¼-inch fall per foot for the piping installed. Divide the length of the offset by 4 to get the correct answer.

4-18 According to Code, a transition fitting for copper (DWV) pipe to clay pipe for a building sewer may be a:

(A) flexible joint   (C) expansion joint

(B) contraction joint   (D) tapered joint

Standard Plumbing Code	Uniform Plumbing Code
Answer: A	**Answer: A**
Code response: Flexible coupling may be used to adapt copper (DWV) to clay sewer pipe having the same O.D.	Code response: Flexible compression joints may be used to adapt copper (DWV) to clay sewer piping.

4-19 When joining threaded piping and fittings together, the Code requires that an approved joint compound be applied:

(A) to male threads only   (C) to pipe threads and to the inside of fitting threads

(B) to female threads only   (D) to the inside of fitting threads only

Standard Plumbing Code	Uniform Plumbing Code
Answer: A	**Answer: A**
Code response: Pipe joint material shall be used only on male threads.	Code response: When pipe joint material is used, it shall be applied only to male threads.

4-20 Of the following water piping joints, the type that is Code-approved for installation in the building water distribution system is:

(A) caulked joints   (C) elastomeric gasketed joints

(B) joints protected with packing additives   (D) soldered joints

Standard Plumbing Code	Uniform Plumbing Code
Answer: D	**Answer: D**

Note: Answers "A" and "C" are Code-approved for **outside** water piping joints. Answer "B" refers to a leak-sealing additive and is prohibited by Code. The only correct answer is "D."

Q 4-21 Joints to be soldered in a copper water piping system, according to Code, must be properly:

(A) tinned
(B) fluxed with an approved corrosive flux

(C) fluxed with an approved noncorrosive flux
(D) cleaned by mechanical means

Standard Plumbing Code

Uniform Plumbing Code

Answer: C
Code response: Soldered joints for copper tubing shall be properly fluxed with an approved noncorrosive flux.

Answer: C
Code response: The joints in copper tubing shall be properly fluxed with an approved noncorrosive flux.

Q 4-22 In a water piping system, a proper flaring tool is permitted by Code to make flared joints for:

(A) type "L" flexible copper tubing
(B) type "L" rigid copper tubing

(C) CPVC plastic piping
(D) lead pipe not exceeding 1 inch in diameter

Standard Plumbing Code

Uniform Plumbing Code

Answer: A
Code response: Flared joints for soft, tempered copper water tubing shall be expanded with a proper flaring tool.

Answer: A
Code response: Flared joints for soft copper water tubing shall be made with a proper flaring tool.

Q 4-23 Connection between floor-mounted plumbing fixtures and drainage pipes shall be made, according to Code, with:

(A) setting seals
(B) flanges

(C) lead and oakum
(D) approved gaskets

Standard Plumbing Code

Uniform Plumbing Code

Answer: B
Code response: Fixture connection between drainage pipes and approved floor outlet plumbing fixtures shall be made with approved flanges.

Answer: B
Code response: Fixture connection between drainage pipes and approved floor outlet plumbing fixtures shall be made with approved flanges.

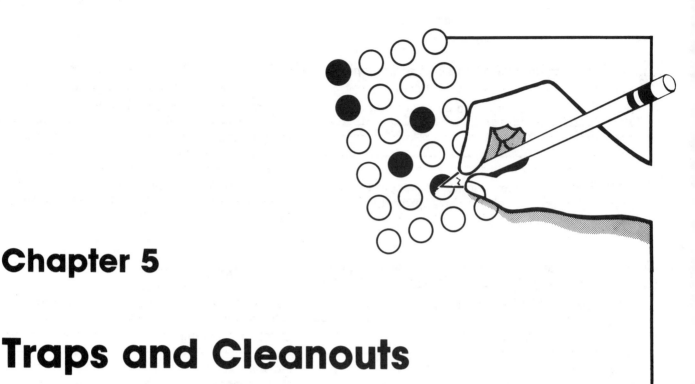

Chapter 5

Traps and Cleanouts

Traps

Before sanitary systems were designed with a venting system to protect the fixture traps, it was common for the water seal in these traps to be broken by the action of back pressure, or siphonage, or both. Rats traveled freely from one building to another. Decomposing sewage in the sewage collection system generated gas and offensive odors which were released into the building through the fixtures. Health department officials recognized that these conditions were a serious health menace to people living in fast-growing cities and towns. They required that a building trap be placed on each building drainage line.

Building traps at that time provided a secondary safeguard to keep rats, vermin, sewer gas and odors out of buildings. Building traps were considered a necessity until the modernization of collection, drainage and venting systems. Today, most model codes don't require (and actually prohibit) the installation of a building trap in a building drainage line. *Verify this in your local code.*

Today's standards require that plumbing fixtures connected directly to the sanitary drainage system be equipped with a water seal trap. The design and construction of this trap provide a liquid seal without materially affecting the flow of sewage or other waste liquids through it.

Because of the trap's unique importance as a protective health measure, the code has placed many restrictions and limitations on its use.

Cleanouts

Before cleanouts were required on drainage piping, the plumber had to cut a hole in any blocked drainage pipe. Obstructions were removed by inserting a cleaning cable through the hole. He then patched the hole with a cement mixture or other impervious material. The patch jobs often deteriorated as time passed, and allowed raw sewage to seep into the ground. This caused a health hazard for the building's occupants and neighbors.

Over the years, cleanouts have become an essential part of the drainage system. Today's model codes clearly outline the location, distance between cleanouts, size, and many other requirements. The questions in this chapter will test your knowledge of requirements for traps and cleanouts.

Q 5-1 By Code requirements, a trap depending on movable parts to retain its seal may:

(A) be used for fixtures with clear water waste only

(B) not be used under any circumstances

(C) be used if first approved by the Plumbing Official

(D) be used when approved by ASME Standards

Standard Plumbing Code	Uniform Plumbing Code
Answer: B Code response: A trap which depends upon the action of movable parts to retain its seal shall not be used.	**Answer: B** Code response: No trap depending for its seal upon movable parts shall be used.

Q 5-2 According to the Code, cleanouts on the seal of a trap are prohibited on:

(A) residential lavatory traps

(B) commercial sink traps

(C) stall shower traps

(D) barber shop sink traps

Standard Plumbing Code	Uniform Plumbing Code
Answer: C Code response: Trap cleanouts shall not be provided on the seal of a fixture trap that is not readily accessible for cleaning purposes.	**Answer: C** Code response: Each cleanout shall be installed, readily accessible and located to serve its intended purpose.

Note: Shower traps are not considered readily accessible. Codes generally prohibit the use of cleanouts on this type trap.

Q 5-3 According to the Code, the smallest trap that may be used for a floor drain is:

(A) 2 inches

(B) 2½ inches

(C) 3 inches

(D) 4 inches

Standard Plumbing Code	Uniform Plumbing Code
Answer: A Code response: Where floor drains are installed below a basement floor, the floor drain trap shall be not less than 2 inches in diameter.	**Answer: A** Code response: Floor drain traps shall be sized to serve efficiently the purpose for which they are intended. Minimum size floor drain listed is 2 inches in diameter.

Q 5-4 According to the Code, 3-inch cleanouts must be accessible and have a minimum clearance (for cleaning purposes) of no less than:

(A) 6 inches

(B) 12 inches

(C) 18 inches

(D) 24 inches

Standard Plumbing Code	Uniform Plumbing Code
Answer: C Code response: 3-inch cleanouts in piping shall be so installed as to provide a clearance of not less than 18 inches.	**Answer: C** Code response: 3-inch cleanouts in piping shall have a clearance of not less than 18 inches.

5-5 According to the Code, line cleanouts should be so installed and arranged that:

(A) only special cleaning tools may be used

(B) special cleaning tools may not be required

(C) grease-dissolving fluids may be easily added

(D) upstream and downstream rodding is possible

Standard Plumbing Code	Uniform Plumbing Code
Answer: D Code response: Line cleanouts which may be rodded both ways **shall** be used **whenever possible.**	**Answer: D** Code response: An approved type of two-way cleanout fitting **may be used.**

Note: Both codes take a permissive attitude toward required line cleanouts. The answer should be "D."

5-6 According to the Code, the integrity of a trap seal is protected by:

(A) proper venting design

(B) having a trap seal of more than 4 inches

(C) having slip joints on both sides of the trap seal

(D) having slip joints on the inlet side of the trap seal only

Standard Plumbing Code	Uniform Plumbing Code
Answer: A Code response: The plumbing system shall be provided with a system of vent piping so that under normal use, the seal of a fixture trap shall not be subjected to siphonage and back pressure.	**Answer: A** Code response: Each plumbing fixture trap shall be protected against siphonage and back pressure by means of vent pipes installed in accordance with this Code.

5-7 According to Code, except when deeper seals are required for interceptors, a fixture trap must have a water seal of between:

(A) 1 and 2 inches

(B) 1 and 3 inches

(C) 2 and 3 inches

(D) 2 and 4 inches

Standard Plumbing Code	Uniform Plumbing Code
Answer: D Code response: Each fixture trap shall have a water seal of not less than 2 inches and not more than 4 inches.	**Answer: D** Code response: Each fixture trap shall have a water seal of not less than 2 inches and not more than 4 inches.

5-8 The one trap that is prohibited by Code is:

(A) one that is not set level

(B) a drum trap

(C) a trap with interior partitions

(D) a trap having less than a 3-inch water seal

Standard Plumbing Code	Uniform Plumbing Code
Answer: C Code response: Partitioned traps are prohibited.	**Answer: C** Code response: No form of trap having a seal that depends upon the action of interior partitions shall be used.

5-9 The Code requires that fixture traps for a lavatory:

(A) be self-cleaning

(B) have a J-bend

(C) have a brass cleanout plug for cleaning purposes

(D) be designed to avoid fungus growth

Standard Plumbing Code

Uniform Plumbing Code

Answer: A
Code response: Fixture traps shall be self-cleaning.

Answer: A
Code response: Each trap shall be self-cleaning.

5-10 The Code requires that a cleanout be installed on each 4-inch horizontal drainage pipe:

(A) at the downstream end of the pipe

(B) at the upstream end of the pipe

(C) no more than 50 feet apart

(D) no less than 100 feet apart

Standard Plumbing Code

Uniform Plumbing Code

Answer: B
Code response: Each horizontal drainage pipe shall be provided with a cleanout at the upstream end of the pipe.

Answer: B
Code response: Each horizontal drainage pipe shall be provided with a cleanout at its upper terminal.

5-11 When cleanouts are required by Code in piping installed under a floor slab, they:

(A) may be omitted if approved by the Plumbing Official

(B) may be smaller than normally required by Code

(C) must be provided with a means of access

(D) must be extended to the outside of the building

Standard Plumbing Code

Uniform Plumbing Code

Answer: D
Code response: Cleanouts on piping under a floor slab shall be extended to the outside of the building.

Answer: D
Code response: Cleanouts in underfloor piping shall be extended outside the building.

Q 5-12 Look at Figure 5-1. There is a maximum allowable vertical drop from a fixture waste outlet to the trap water seal. To meet the Code, the distance "X" must not exceed _____ inches.

Figure 5-1

(A) 12

(B) 18

(C) 24

(D) 28

Standard Plumbing Code

Answer: C
Code response: The vertical distance from the fixture outlet to the trap weir shall not exceed 24 inches.

Uniform Plumbing Code

Answer: C
Code response: The vertical distance between a fixture outlet and the trap weir shall not exceed 24 inches.

Q 5-13 The fitting shown in Figure 5-2 is known as a countersunk cleanout plug. According to Code, it must be used:

Figure 5-2

(A) where cleanouts terminate with finished outside walls

(B) where cleanouts terminate with finished floors

(C) when installed in a kitchen sink cabinet

(D) at the base of exposed waste stacks

Standard Plumbing Code

Answer: B
Code response: Countersunk cleanout heads shall be used where raised heads may cause a hazard.

Uniform Plumbing Code

Answer: B
Code response: Countersunk cleanout plugs shall be used where raised cleanout plugs may cause a hazard.

5-14 According to Code, all of the following types of traps are prohibited for use in a drainage system, except:

(A) the bell trap
(B) the crown-vented trap

(C) the P-trap
(D) the full "S" trap

Standard Plumbing Code

Uniform Plumbing Code

Answer: C
Code response: Prohibited traps are "S" traps, bell traps, and crown-vented traps.

Answer: C
Code response: Traps prohibited are full "S" traps, bell traps, and crown-vented traps.

5-15 A plumbing inspector, while making a final inspection, noticed that a laundry trap was tilted downward toward the drain pipe at an angle of about 10 degrees. According to Code, the inspector would turn the job down because:

(A) the depth of the trap seal would be reduced

(B) the trap would retain solids

(C) the trap might form an "S" trap, which is prohibited by Code
(D) the trap is not set level to its water seal

Standard Plumbing Code

Uniform Plumbing Code

Answer: D
Code response: Traps shall be set level with respect to their water seal.

Answer: D
Code response: Traps shall be set true with respect to their water seal.

5-16 According to the Code, 2-inch cleanouts must be accessible for cleaning purposes and have a minimum clearance of:

(A)   6 inches
(B) 12 inches

(C) 18 inches
(D) 24 inches

Standard Plumbing Code

Uniform Plumbing Code

Answer: B
Code response: Cleanouts smaller than 3 inches shall have a clearance no less than 12 inches.

Answer: B
Code response: Cleanouts 2 inches or less in size shall have a clearance of not less than 12 inches.

5-17 Several adjacent lavatories are installed on the same wall in the same room. According to Code, _____ lavatories may use a single trap.

(A) 2
(B) 3

(C) 4
(D) 5

Standard Plumbing Code

Uniform Plumbing Code

Answer: B
Code response: Provided that 1 trap may be installed for a set of not more than 3 lavatories immediately adjacent to each other and in the same room.

Answer: B
Code response: It is provided that 1 trap may serve a set of not more than 3 lavatories immediately adjacent to each other and in the same room.

Q 5-18 According to Code, all plumbing fixtures, except those having _____, shall be individually trapped.

(A) continuous waste
(B) an end outlet waste

(C) integral traps
(D) water seals not approved by Code

Standard Plumbing Code

Uniform Plumbing Code

Answer: C
Code response: Plumbing fixtures, excepting those having integral traps, shall be individually trapped.

Answer: C
Code response: Each plumbing fixture, excepting those having integral traps, shall be individually trapped.

Q 5-19 In Figure 5-3, the maximum distance "X" allowed by Code between the fixtures is _____ inches.

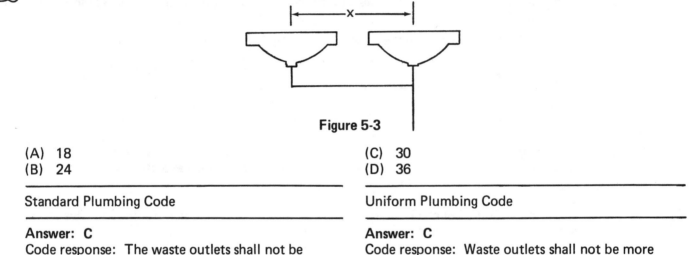

Figure 5-3

(A) 18
(B) 24

(C) 30
(D) 36

Standard Plumbing Code

Uniform Plumbing Code

Answer: C
Code response: The waste outlets shall not be more than 30 inches apart.

Answer: C
Code response: Waste outlets shall not be more than 30 inches apart.

Q 5-20 Look at Figure 5-4. Three single-compartment sinks are installed at the same level and on the same wall. The plumber intends to serve the 3 sinks with 1 trap. According to Code, he must install the trap under sink _____.

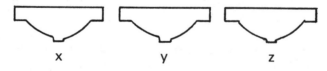

Figure 5-4

(A) X
(B) Y

(C) Z
(D) any of the above

Standard Plumbing Code

Uniform Plumbing Code

Answer: B
Code response: 3 single-compartment sinks may use 1 trap, providing it is centrally located.

Answer: B
Code response: 3 single-compartment sinks may use 1 trap, providing it is centrally located.

Q 5-21 The Code permits connecting a 1½-inch bar sink trap to a fixture drain, provided it is:

(A) not less than 1¼ inches I.D.   (C) not less than 1½ inches I.D.
(B) not less than 1¼ inches O.D.   (D) not less than 1½ inches O.D.

Standard Plumbing Code   Uniform Plumbing Code

Answer: C   **Answer: C**
Code response: A fixture trap shall not be larger than the fixture drain (trap arm) to which it is connected.   Code response: A trap shall be the same size as the trap arm (fixture drain) to which it is connected.

Q 5-22 According to Code, a trap which is integral with the fixture must:

(A) be no smaller than 2 inches   (C) be self-sealing
(B) be no larger than 3 inches   (D) have a smooth waterway

Standard Plumbing Code   Uniform Plumbing Code

Answer: D   **Answer: D**
Code response: A trap which is integral shall have a smooth waterway.   Code response: Every trap shall have a smooth waterway.

Q 5-23 According to Code, the size of a fixture trap must be:

(A) sufficient to drain the fixture quickly   (C) no larger than the fixture tail piece
(B) accessible   (D) no less than 17-gauge metal

Standard Plumbing Code   Uniform Plumbing Code

Answer: A   **Answer: A**
Code response: The size of a given fixture trap shall be sufficient to drain the fixture rapidly.   Code response: The size of a given fixture trap shall be sufficient to drain the fixture rapidly.

Q 5-24 The Code specifies that a drum trap must have a water seal of not less than _____ inch(es).

(A) 1   (C) 3
(B) 2   (D) 4

Standard Plumbing Code   Uniform Plumbing Code

Answer: B   **Answer: B**
Code response: Drum traps shall have a water seal of not less than 2 inches.   Code response: Plumbing fixture traps shall have a water seal of not less than 2 inches.

5-25 According to Code, floor drains subject to backflow and connected to a 3-inch cast-iron P-trap must be:

(A) indirectly connected to the drainage system

(B) directly connected to the drainage system

(C) provided with a backwater valve

(D) installed with a greater pitch to drain faster

Standard Plumbing Code	Uniform Plumbing Code
Answer: C Code response: Floor drains subject to backflow shall be provided with a backwater valve.	**Answer: C** Code response: Floor drains when subject to backflow shall be equipped with an approved backwater valve.

5-26 The Code requires that floor drains installed in areas of infrequent use and connected directly to the drainage system be provided with a:

(A) basket and strainer

(B) backwater valve

(C) removable plug

(D) trap seal protection device

Standard Plumbing Code	Uniform Plumbing Code
Answer: D Code response: Floor drain trap seals subject to evaporation shall be provided with an automatic priming device.	**Answer: D** Code response: Floor drains subject to infrequent use shall be provided with a means of maintaining their water seals.

5-27 To meet the Code, commercial food waste disposal units must be connected:

(A) directly to the building waste line

(B) to a separate trap

(C) to a floor-mounted grease trap

(D) to a 2-inch pot sink trap

Standard Plumbing Code	Uniform Plumbing Code
Answer: B Code response: Each waste line from a commercial type food grinder shall be trapped.	**Answer: B** Code response: Commercial food waste disposal units shall be connected to a separate trap.

5-28 The Code allows slip joints for a tubular P-trap to be used:

(A) on the inlet side of a fixture trap

(B) on the outlet side of a fixture trap

(C) on the trap seal of a trap

(D) in all of the above locations

Standard Plumbing Code	Uniform Plumbing Code
Answer: D Code response: Slip joints may be used on both sides of the trap and in the trap seal.	**Answer: D** Code response: No more than 1 approved slip joint fitting may be used on the outlet side of the trap.

Note: Check local Code for your particular specifications.

5-29 According to Code, the pipe between the trap weir and the vent pipe is known as a:

(A) trap arm

(B) fixture branch

(C) fixture tail piece

(D) fixture waste pipe

Standard Plumbing Code

Answer: A
Code response: A fixture drain (trap arm) is the drain pipe from the trap weir to the inner edge of the vent.

Uniform Plumbing Code

Answer: A
Code response: A fixture trap arm (fixture drain) is the drain pipe from the trap weir to the inner edge of the vent.

5-30 According to Code, traps for floor drains must be so constructed as to:

(A) avoid the deposit of solids

(B) maintain their water seal

(C) avoid evaporation

(D) be readily cleaned

Standard Plumbing Code

Answer: D
Code response: Floor drains shall connect into a trap so constructed that it can be readily cleaned.

Uniform Plumbing Code

Answer: D
Code response: Floor drains shall connect into a trap so constructed that it can be readily cleaned.

5-31 According to the Code, a ____ shall not be larger than the fixture drain to which it is connected.

(A) cleanout

(B) trap

(C) branch drain

(D) relief vent

Standard Plumbing Code

Answer: B
Code response: A trap shall not be larger than the fixture drain to which it is connected.

Uniform Plumbing Code

Answer: B
Code response: The trap shall be the same size as the trap arm (fixture drain) to which it is connected.

5-32 Code requires that the water seal of a fixture trap be not more than ____ inches.

(A) 3

(B) 4

(C) 5

(D) 6

Standard Plumbing Code

Answer: B
Code response: A fixture trap shall have a water seal of not more than 4 inches.

Uniform Plumbing Code

Answer: B
Code response: Each fixture trap shall have a water seal not more than 4 inches.

Q 5-33 When the Administrative Authority allows manholes to be installed on a building sewer, Code requires that such manholes be no less than _____ feet apart.

(A) 100
(B) 150

(C) 200
(D) 300

Standard Plumbing Code | Uniform Plumbing Code

Answer: D
Code response: For building sewers 8 inches and larger, manholes shall be provided at intervals not exceeding 300 feet.

Answer: D
Code response: The maximum distance between manholes shall not exceed 300 feet.

Q 5-34 According to Code, a cleanout must be installed:

(A) on each drainage pipe

(B) at the base of each stack

(C) near the junction of the building drain and building sewer
(D) at the property line and brought to grade

Standard Plumbing Code | Uniform Plumbing Code

Answer: C
Code response: Each building drain shall have installed a cleanout near the junction of the building drain and building sewer.

Answer: C
Code response: Cleanouts shall be placed inside the building near the connection between the building drain and the building sewer.

Q 5-35 According to Code, the approved trap that is most successful in staying clog-free is the:

(A) "S" trap
(B) trap having movable parts

(C) bell trap
(D) P-trap

Standard Plumbing Code | Uniform Plumbing Code

Answer: D
Code response: Fixture traps shall be self-cleaning.

Answer: D
Code response: Fixture traps shall be self-cleaning.

Note: The P-trap is the only Code-approved trap listed above.

Q 5-36 Code states that the minimum size trap for a standard service sink be _____ inches.

(A) 1½
(B) 2

(C) 2½
(D) 3

Standard Plumbing Code | Uniform Plumbing Code

Answer: B
Code response: The minimum size trap for a service sink (P-trap) is 2 inches.

Answer: B
Code response: The minimum size trap for a service sink is 2 inches.

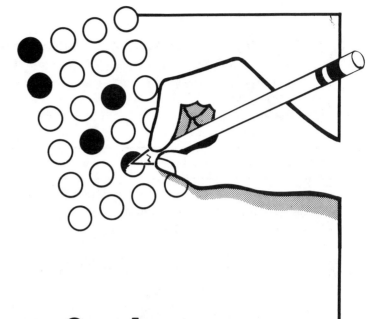

Chapter 6

Sanitary Drainage Systems

The private sanitary drainage system is the most basic part of the plumber's work. It includes:

1) all the pipes installed within the wall line of a building and on private property for the purpose of receiving liquid waste or other waste substances (whether in suspension or in solution).

2) the pipes which convey this waste to a public sewer or other private, approved sewage disposal system.

The drainage system must be installed so that it isn't a health or safety hazard to any individual or to the general public. Although details of plumbing systems may vary, the basic principles of sanitation and safety remain the same, regardless of the code adopted in your area.

The type and number of plumbing fixtures determine the *fixture units* necessary for sizing the various pipes in a drainage system. The code regulates the type of materials and fittings, supports, changes of direction, location for cleanouts, distances of fixture traps from vent pipes, minimum slopes for horizontal pipes, installation methods, and much more. Although the sanitary drainage system is considered the heart of the plumbing system, most experts agree that this is the most complex, misunderstood, and misinterpreted section of the code.

Most of the questions asked (and most isometric drawings, when required) on the journeyman and master examinations cover the drainage system.

The questions in this chapter will test your knowledge of this subject. Since so many exam questions cover this part of the code, devote a large portion of your study time to reviewing this chapter. Don't go into the exam with less than an excellent understanding of drainage systems.

6-1 According to the Code, the private sanitary drainage system does **not** include:

(A) the pipes installed within the wall line of a building to receive waste substances in suspension

(B) the building sewer

(C) the pipe which conveys waste substances from a garbage-can wash

(D) the pipes installed to convey waste substances from area drains

---

Standard Plumbing Code | Uniform Plumbing Code

Answer: D
Code response: It is prohibited to drain storm water into sewers intended for sewage only.

Answer: D
Code response: Rain water drainage shall not be connected to the drainage system.

 6-2 According to the Code, the drainage system is sized by:

(A) the maximum fixture unit load

(B) the minimum fixture unit load

(C) the number of fixtures connected thereto

(D) a state-registered engineer

Standard Plumbing Code | Uniform Plumbing Code

Answer: A
Code response: The maximum number of fixture units that may be connected to a given size of piping within the drainage system.

Answer: A
Code response: The maximum number of fixture units allowed on any vertical or horizontal drainage pipe, building drain or building sewer of a given size.

 6-3 According to the Code, when calculating the fixture unit load for a 30-unit apartment building having 30 bathtubs with shower heads, the shower heads will:

(A) increase the total load by 15 fixture units

(B) increase the total load by 30 fixture units

(C) not increase the total fixture unit load value

(D) increase the total load by ½ fixture unit when in use

Standard Plumbing Code | Uniform Plumbing Code

Answer: C
Code response: A shower head over a bathtub will **not** increase the fixture unit load.

Answer: C
Code response: Does not address this particular question.

Note: A shower head is already calculated in the fixture unit value of a bathtub. Answer "C" is correct, even though the Uniform Plumbing Code does not address this particular question.

 6-4 According to the Code, cast-iron no-hub pipe and fittings used in the installation of drainage and vent systems shall have the approved standards established by the:

(A) American National Standards Institute, Inc. (ANSI)

(B) American Society of Sanitary Engineering (ASSE)

(C) American Society for Testing Materials (ASTM)

(D) Cast-Iron Soil Pipe Institute (CISPI)

Standard Plumbing Code | Uniform Plumbing Code

Answer: D
Code response: Standards given in Table 500 apply to the specific use of certain materials approved as each applies to the drainage system.

Answer: D
Code response: Standards given in Table A apply to the specific use of certain materials approved as each applies to the drainage system.

Q 6-5 According to the Code, a building sewer installed in filled or unstable ground shall be of:

(A) DWV schedule 80

(B) DWV copper—type L

(C) cast-iron pipe

(D) any approved metallic piping and fittings

Standard Plumbing Code	Uniform Plumbing Code

Answer: C

Code response: A building sewer, when installed in filled or unstable ground, shall be of cast-iron pipe.

Answer: ____

Code response: Does not address this particular type of installation.

Note: If your primary Code is the Uniform Plumbing Code, you probably don't need to concern yourself with this question. But check amendments in your particular area for confirmation.

Q 6-6 Fittings used in a PVC-type DWV drainage system shall:

(A) not be used where combustible construction is allowed

(B) conform to the type of pipe used

(C) be schedule 80 only

(D) be of the recessed type

Standard Plumbing Code	Uniform Plumbing Code

Answer: B

Code response: Fittings on the drainage system shall conform to the type of pipe used.

Answer: B

Code response: Drainage fittings shall conform to the type of pipe used.

Q 6-7 To meet Code requirements, a 2-inch-diameter waste pipe installed horizontally must fall a minimum of ____ inches every 35 feet.

(A) 5-1/8

(B) 6-5/8

(C) 7-1/4

(D) 8-3/4

Standard Plumbing Code	Uniform Plumbing Code

Answer: D

Code response: Horizontal drainage piping of 2-inch diameter shall have a minimum fall of ¼-inch per foot.

Answer: D

Code response: 2-inch horizontal drainage piping shall be run in a uniform slope of not less than ¼-inch per foot.

Note: Here's the solution: 35.0 feet x 0.25 inch per foot = 8.75 inches.

Q 6-8 According to the Code, ____ must discharge their waste through a grease interceptor.

(A) garbage-can washers

(B) commercial food grinders

(C) garage floor drains

(D) acid waste systems

Standard Plumbing Code	Uniform Plumbing Code

Answer: A

Code response: Garbage-can washers shall be connected through a grease interceptor.

Answer: ____

Code response: Not addressed. The local Administrative Authority has sole jurisdiction.

Note: Garbage-can washers are becoming obsolete in restaurants. However, where they are permitted by Code, the waste is usually required to go through a grease interceptor.

Q 6-9 According to Code, horizontal combination waste-and-vent sanitary systems are limited to:

(A) water closets only

(B) floor drains only

(C) floor sinks only

(D) fixtures not adjacent to walls or partitions

Standard Plumbing Code

Uniform Plumbing Code

Answer: D

Code response: Horizontal combination waste-and-vent sanitary systems are limited to **fixtures not adjacent to walls or partitions** and are limited to sinks, dishwashers, indirect waste receptors, floor drains or similar fixtures.

Answer: D

Code response: Horizontal combination waste-and-vent sanitary systems are limited to **fixtures not adjacent to walls or partitions** and are intended primarily for extensive floor or shower drain installations, floor sinks and demonstration or work tables having waste outlets or for similar applications.

Q 6-10 According to the Code, the smallest pipe diameter for a soil stack that carries no waste from urinals or bedpan washers is _____ inches.

(A) 2

(B) 2½

(C) 3

(D) 4

Standard Plumbing Code

Uniform Plumbing Code

Answer: C

Code response: No water closet shall discharge into a stack unless it has a minimum diameter of 3 inches.

Answer: C

Code response: No water closet shall discharge into a stack of less than 3 inches in diameter.

Q 6-11 A 4-inch-diameter building drain installed horizontally must have a minimum of _____ fall per foot to meet the Code.

(A) 1/16-inch

(B) 1/8-inch

(C) 1/4-inch

(D) 1/2-inch

Standard Plumbing Code

Uniform Plumbing Code

Answer: B

Code response: Horizontal building drains 4 inches in diameter or larger shall have no less than 1/8-inch fall per foot.

Answer: B

Code response: Horizontal drainage piping 4 inches or larger in diameter may have a slope of not less than 1/8-inch per foot.

Q 6-12 Fixture unit equivalents not commonly listed in fixture unit tables, according to Code, must be based on the:

(A) type of fixture

(B) trap size

(C) location of fixture

(D) discharge capacity of fixture

Standard Plumbing Code

Uniform Plumbing Code

Answer: B

Code response: The unit equivalent of fixtures not listed shall be based on the fixture trap size.

Answer: D

Code response: The unit equivalent of plumbing fixtures shall be based on the rated discharge capacity in gpm.

Note: Check local Code for fixture unit values for special plumbing fixtures. In sizing a drainage system, the method of calculating special plumbing fixture load factors varies in these two Codes.

6-13 Of the following drainage piping materials used within a building, the one prohibited by Code for underground installation is:

(A) DWV copper piping
(B) lead piping

(C) steel piping
(D) brass piping

Standard Plumbing Code

Uniform Plumbing Code

Answer: C
Code response: None.

Answer: C
Code response: No galvanized wrought-iron or galvanized steel pipe shall be installed underground in a drainage system.

Note: The Standard Plumbing Code does not specifically answer this question, but it does **not** list galvanized wrought-iron or steel pipe as approved material for underground installation. The correct answer is "C."

6-14 Fittings used within a drainage system, according to Code, must:

(A) be constructed of cast iron
(B) have a drainage pattern from horizontal to horizontal

(C) have a smooth interior waterway
(D) have a drainage pattern of not less than 45 degrees

Standard Plumbing Code

Uniform Plumbing Code

Answer: C
Code response: Fittings which offer abnormal obstruction to flow shall not be permitted in a drainage system.

Answer: C
Code response: Drainage fittings shall have a smooth interior waterway.

6-15 The minimum size vertical waste pipe required by Code for a service sink having a 2-inch diameter trap is:

(A) 1½ inches
(B) 2 inches

(C) 2½ inches
(D) 3 inches

Standard Plumbing Code

Uniform Plumbing Code

Answer: B
Code response: A trap shall not be larger than the waste pipe to which it is connected.

Answer: B
Code response: A trap shall not be larger than the waste pipe to which it is connected.

6-16 Assume that waste is discharged from a building at a rate of 630 gallons per minute. The rate of flow in terms of fixture units is:

(A) 62
(B) 76

(C) 84
(D) 108

Standard Plumbing Code

Uniform Plumbing Code

Answer: C
Code response: A lavatory is considered 1 fixture unit. A fixture unit flow rate shall be deemed 7.5 gallons of water per minute.

Answer: C
Code response: A lavatory is considered 1 fixture unit. A fixture unit flow rate shall be deemed 7.5 gallons of water per minute.

Note: Here's the solution: 630 gpm ÷ 7.5 gpm = 84

Q 6-17 Of the following factors, the one that is <u>not</u> required by Code for determining the size of a drainage pipe is:

(A)   the grade of the piping
(B)   the type of piping material used

(C)   the type of fixtures used
(D)   the length of run

Standard Plumbing Code

Uniform Plumbing Code

Answer: B
Code response: This Code does not address this particular question as presented.

Answer: B
Code response: This Code does not address this particular question as presented.

Note: There are several types of piping material approved by Code for use in a drainage system. The **type** of piping material does **not** determine the size of the piping in a drainage system.

Q 6-18 The minimum sizes of horizontal drainage piping, according to Code, are determined by:

(A)   the waste outlet size of fixtures connected thereto
(B)   the number and type of fixtures connected thereto

(C)   the length of run and pitch per foot

(D)   the total of all fixture units connected thereto

Standard Plumbing Code

Uniform Plumbing Code

Answer: D
Code response: The minimum sizes of horizontal drainage piping shall be determined from the total of all fixture units connected thereto.

Answer: D
Code response: The minimum sizes of horizontal drainage piping shall be determined from the total of all fixture units connected thereto.

Q 6-19 In the case of vertical drainage piping, the minimum sizes, according to Code, are determined by the total of all fixture units connected thereto, and by:

(A)   any offsets in the stack
(B)   their length

(C)   the minimum vent piping size required
(D)   the largest fixture opening at the highest level

Standard Plumbing Code

Uniform Plumbing Code

Answer: B
Code response: The minimum sizes of vertical drainage piping shall be determined from the total of all fixture units connected thereto and, in addition, in accordance with their length.

Answer: B
Code response: The minimum sizes of vertical drainage piping shall be determined from the total of all fixture units connected thereto and, additionally, in accordance with their length.

6-20 Where there is a continuous flow from an air-conditioning unit into a building drainage system, each gpm of flow, according to Code, must equal ____ fixture unit(s).

(A) ½

(B) 1

(C) 1½

(D) 2

Standard Plumbing Code

Uniform Plumbing Code

Answer: D

Code response: For a continuous flow into a drainage system, such as from air-conditioning equipment, 2 fixture units shall be allowed for each gallon per minute of flow.

Answer: D

Code response: For a continuous flow into a drainage system, such as from air-conditioning equipment, 2 fixture units shall be allowed for each gallon per minute of flow.

6-21 Assume the flow from a pump ejector in a building is 66 gallons per minute. According to Code, this would add a total of ____ fixture units to the building drainage system.

(A) 123

(B) 128

(C) 132

(D) 137

Standard Plumbing Code

Uniform Plumbing Code

Answer: C

Code response: A pump ejector shall be allowed 2 fixture units (FU) for each gallon per minute of flow.

Note: Here's the solution: 66 gpm x 2 FU = 132 FU

Answer: C

Code response: A pump ejector shall be allowed 2 fixture units (FU) for each gallon per minute of flow.

6-22 Any part of a piping system that extends horizontally at an approved grade, with or without vertical extensions from the main to plumbing fixtures, is called a ____ in the Code.

(A) lateral

(B) continuous pipe

(C) branch drain

(D) branch

Standard Plumbing Code

Uniform Plumbing Code

Answer: D

Code response: A branch is any part of the piping system other than a main, riser, or stack.

Answer: D

Code response: A branch is any part of the piping system other than a main, riser, or stack.

6-23 Any part of a drainage system that extends horizontally at a slope from a soil or waste stack, with or without lateral or vertical extensions, on two or more floors, is defined by Code as a:

(A) horizontal branch

(B) horizontal pipe

(C) fixture drain

(D) combination fixture drain

Standard Plumbing Code

Uniform Plumbing Code

Answer: A

Code response: A horizontal branch is a drain pipe extending laterally from a soil or waste stack, with or without vertical sections or branches.

Answer: A

Code response: A horizontal branch is a drain pipe extending laterally from a soil or waste stack, with or without vertical sections or branches.

6-24 To connect a 3-inch vertical stack to a 4-inch horizontal drainage line, the acceptable fitting, by Code, is a:

(A)   4-inch by 4-inch by 3-inch sanitary tee

(B)   4-inch by 3-inch by 4-inch combination

(C)   4-inch by 3-inch by 4-inch sanitary tee

(D)   4-inch by 4-inch by 3-inch combination

Standard Plumbing Code	Uniform Plumbing Code
Answer: D Code response: Vertical stacks connecting to a horizontal drainage line shall be made by the appropriate use of a combination of (or equivalent) fittings to provide a 45-degree branch.	**Answer: D** Code response: Vertical drainage lines connecting to horizontal drainage lines shall enter through 45-degree branches.

6-25 The fitting shown in Figure 6-1, according to Code, may be used in a drainage system:

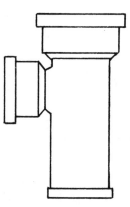

Figure 6-1

(A)   to provide a horizontal-to-vertical change of direction

(B)   to provide a horizontal-to-horizontal change of direction

(C)   both "A" and "B"

(D)   none of the above

Standard Plumbing Code	Uniform Plumbing Code
Answer: D Code response: A straight tee branch shall not be installed as a drainage fitting in a plumbing drainage system.	**Answer: D** Code response: No single or double tee branch shall be used as a drainage fitting.

6-26 The fitting shown in Figure 6-2, according to Code, may be used in a drainage system when the direction of flow is from the:

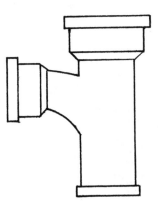

Figure 6-2

(A) horizontal to vertical
(B) horizontal to horizontal

(C) vertical to horizontal
(D) none of the above

Standard Plumbing Code

Uniform Plumbing Code

Answer: A
Code response: Sanitary tees may be used in drainage lines where the direction of flow is from the horizontal to the vertical.

Answer: A
Code response: Sanitary tees are approved for drainage lines when the direction of flow is from the horizontal to the vertical.

6-27 The Code generally requires that piping with a high percentage of silicon be installed to convey:

(A) backwash waste from swimming pools
(B) waste containing acid

(C) waste from industrial plants
(D) waste from hospital operating rooms

Standard Plumbing Code

Uniform Plumbing Code

Answer: B
Code response: Acid waste piping shall be constructed of a high-silicon cast-iron pipe.

Answer: B
Code response: Waste piping receiving the discharge from any fixture in which acid is placed shall be constructed of high-silicon iron pipe.

6-28 A 4-inch sewer pipe, according to Code, must have a fall no less than _____ inches every 50 feet.

(A) 5.5
(B) 6

(C) 6.25
(D) 6.75

Standard Plumbing Code

Uniform Plumbing Code

Answer: C
Code response: Horizontal drainage piping 3 inches and larger shall be installed with a fall of not less than 1/8-inch per foot.

Answer: C
Code response: Horizontal drainage piping 4 inches and larger may have a slope of not less than 1/8-inch per foot.

Note: Remember that 1/8-inch is the same as 0.125 inch. Here's the solution for this question: 50 feet x 0.125-inch per foot = 6.25 feet.

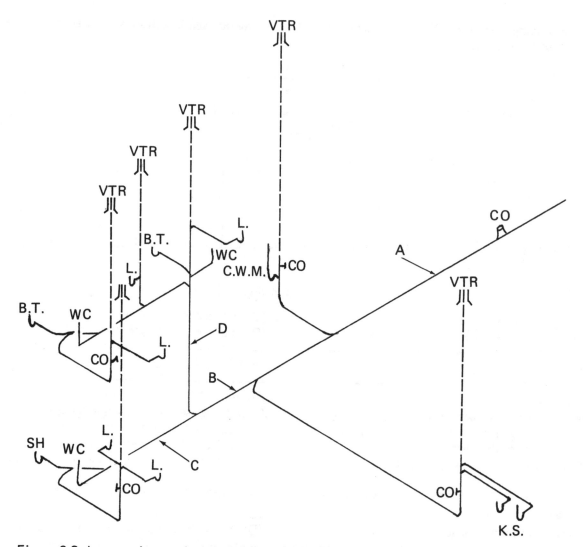

Figure 6-3 shows an isometric drawing for a 3-bath, 2-level residence. The next 4 questions relate to the total fixture units discharging into the drainage system at each listed location, according to Code. The fixtures selected for these questions have the same FU values, according to the 2 Codes. For other types of fixtures, refer to local Code.

Figure 6-3

6-29 Building sewer "A" in Figure 6-3 would have an accumulative fixture unit load value of _____ , according to Code.

(A) 14
(B) 22

(C) 27
(D) 29

Standard Plumbing Code	Uniform Plumbing Code

Answer: C
Code response: Check fixture units per fixture, as listed in the fixture table.

Answer: C
Code response: Check fixture units per fixture, as listed in the fixture table.

Note: Add up the total fixture units for the residence in Figure 6-3. The correct answer is "C."

Q 6-30 Building drain "B" in Figure 6-3 would have an accumulative fixture unit load value of _____ , according to Code.

(A) 13
(B) 16

(C) 19
(D) 22

Standard Plumbing Code

Uniform Plumbing Code

Answer: D
Code response: Check fixture units per fixture, as listed in the fixture table.

Answer: D
Code response: Check fixture units per fixture, as listed in the fixture table.

Note: Count the total number of fixture units up to the listed location "B." The correct answer is "D."

Q 6-31 Building drain "C" in Figure 6-3 would have an accumulative fixture unit load value of _____ , according to Code.

(A) 6
(B) 8

(C) 10
(D) 12

Standard Plumbing Code

Uniform Plumbing Code

Answer: B
Code response: Check fixture units per fixture, as listed in the fixture table.

Answer: B
Code response: Check fixture units per fixture, as listed in the fixture table.

Note: Count the total number of fixture units up to the listed location "C." The correct answer is "B."

Q 6-32 Soil stack "D" in Figure 6-3 would have an accumulative fixture unit load value of _____ , according to Code.

(A) 12
(B) 15

(C) 18
(D) 22

Standard Plumbing Code

Uniform Plumbing Code

Answer: B
Code response: Check fixture units per fixture, as listed in the fixture table.

Answer: B
Code response: Check fixture units per fixture, as listed in the fixture table.

Note: Count the total number of fixture units up to the listed location "D." The correct answer is "B."

Q 6-33 Fittings on threaded pipe installed in a drainage system, according to Code, must be:

(A) of the recessed drainage type
(B) full size in diameter

(C) extra-strength malleable iron
(D) used only where combustible construction is allowed

Standard Plumbing Code

Uniform Plumbing Code

Answer: A
Code response: Fittings on threaded pipe used within a drainage system shall be of the recessed drainage type.

Answer: A
Code response: Drainage fittings on screwed pipe shall be of the recessed drainage type.

Q 6-34 The discharge line from a sewage ejector, according to Code, must connect to the building drain through a:

(A) backwater valve
(B) combination fitting

(C) wye "Y" fitting
(D) 60-degree fitting

Standard Plumbing Code

Uniform Plumbing Code

Answer: C
Code response: Ejector pumps shall connect to a wye "Y" fitting in the building drain.

Answer: C
Code response: The discharge line from ejector pumps shall connect to a horizontal drainage line from the top through a wye "Y" branch fitting.

Q 6-35 The discharge line from a sewage ejector, according to Code, must have a _____ installed.

(A) check valve
(B) gate valve

(C) backwater valve
(D) check valve and gate valve

Standard Plumbing Code

Uniform Plumbing Code

Answer: D
Code response: A check valve and a gate valve shall be installed in the ejector discharge piping.

Answer: D
Code response: The discharge line from an ejector shall be provided with an accessible swing check valve and gate valve.

Q 6-36 Code requires that when subsoil drainage systems are installed, they must discharge:

(A) into an approved sump
(B) through an approved oil-and-sand interceptor

(C) through bucket-type floor drains
(D) in a manner satisfactory to the plumbing inspector

Standard Plumbing Code

Uniform Plumbing Code

Answer: A
Code response: Building drains which cannot be discharged by gravity to the drainage system shall be discharged into an approved sump.

Answer: A
Code response: When subsoil drainage systems are installed, they shall be discharged into an approved sump.

Q 6-37 In geographical areas where temperatures may drop below 32 degrees Fahrenheit, the Code prohibits the installation of soil or waste pipes:

(A) in a wooden partition
(B) below basement floors

(C) in an exterior wall
(D) in a concrete block wall

Standard Plumbing Code

Uniform Plumbing Code

Answer: C
Code response: No soil or waste pipe shall be installed or permitted in outside walls where it may be subjected to freezing temperatures, unless provisions are made to protect such pipe from freezing.

Answer: C
Code response: No soil or waste pipe shall be installed or permitted in an exterior wall, unless provision is made to protect such pipe from freezing.

Q 6-38 Assume a restaurant is serving 150 people at one seating. A full size grease interceptor is required. Of the following, the _____ is prohibited by Code from discharging into the greasy waste system.

(A) garbage-can washer
(B) 3-compartment pot sink

(C) commercial dishwasher
(D) food-waste grinder

Standard Plumbing Code	Uniform Plumbing Code

Answer: D
Code response: Where food-waste grinders are installed, the waste shall not pass through a grease interceptor.

Answer: D
Code response: No food-waste disposal unit shall discharge into any grease interceptor.

Q 6-39 For 2 coffee urns with ¾-inch waste outlets, 2 glass sinks with 1¼-inch waste outlets, 1 cooler box with a ½-inch waste outlet and 2 ice water stations with ½-inch waste outlets, the equivalent fixture unit value, according to Code, is _____ .

(A) 5
(B) 7

(C) 9
(D) 11

Standard Plumbing Code	Uniform Plumbing Code

Answer: B
Code response: The fixture unit value of plumbing fixtures and devices shall be based upon the fixture drain or trap size: 1¼ inches and smaller shall equal 1 fixture unit.

Answer: B
Code response: The unit equivalent of plumbing fixtures and devices shall be based upon the size of trap required: 1¼ inches or less shall equal 1 fixture unit.

Note: For special fixtures and devices not listed in the regular fixture unit table, the above response gives the only waste outlet size that the two Codes agree on. Check local Code to verify fixture units for special plumbing fixtures and devices with larger drains.

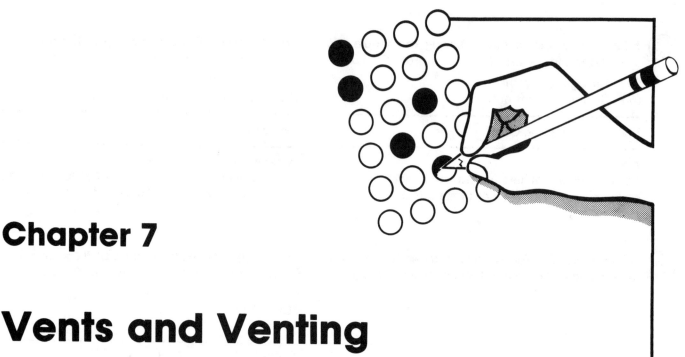

Chapter 7

Vents and Venting

Vent pipes are the unsung heroes of the sanitary drainage system. Proper venting is absolutely essential if the drainage system is to work right.

By the year 1875, it was pretty well established that all vent pipes must extend to the atmosphere above the building roof. Five years later, exact minimum vent pipe sizes were established to serve the various fixture trap sizes.

Vent pipes are sized and arranged to relieve pressure that builds up as water is discharged into the drainage system. The free flow of air within the system keeps back pressure or siphoning action from destroying the fixture trap seal. Vent sizes and the arrangement of vents are determined by the number and type of plumbing fixtures.

Plumbing codes differ on vents and venting requirements, as you'll discover when reading the questions and code responses in this chapter. It's important to be aware of this. Review the venting chapter in your local code so you understand the local requirements for vent sizes and lengths. Use *only* your approved code when answering exam questions on vents and venting.

In spite of the differences between the various codes on vents and venting, most of the information in this section will be useful when taking the exam and when installing vent systems.

7-1 An individual vent pipe that serves a 4-inch waste pipe, according to Code, cannot be smaller in diameter than _____ inches.

(A) 1½

(B) 2

(C) 2½

(D) 3

Standard Plumbing Code	Uniform Plumbing Code

Answer: B

Code response: Individual vent diameters shall not be less than ½ the diameter of the drain served.

Answer: B

Code response: The diameter of an individual vent shall not be less than ½ the diameter of the drain to which it is connected.

7-2 In Figure 7-1, assume that the developed length of the vent piping does not exceed 51 feet. The minimum size of the Code-specified vent piping should be:

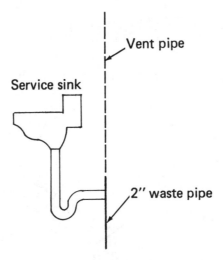

Vent pipe

Service sink

2″ waste pipe

Figure 7-1

(A) 1½ inches

(B) 2 inches

(C) 2½ inches

(D) 3 inches

Standard Plumbing Code	Uniform Plumbing Code

Answer: A

Code response: For 1½-inch individual vertical vent pipes, the Code permits a maximum of 12 fixture units and a developed length of 75 feet.

Answer: A

Code response: For 1½-inch vertical vent pipes, the Code permits a maximum of 8 fixture units and a developed length of 60 feet.

7-3 **To avoid frost or snow closure, the Code requires that vent extensions be no smaller than:**

(A) 2 inches
(B) 2½ inches

(C) 3 inches
(D) 4 inches

Standard Plumbing Code

Uniform Plumbing Code

Answer: C
Code response: Where there is possibility of frost closure, each vent extension through a roof shall be at least 3 inches in diameter.

Answer: C
Code response: Where frost or snow closure is likely to occur, each vent extension through a roof shall be at least 3 inches in diameter.

7-4 **In Figure 7-2, the horizontal portion "X" of the dry vent piping, according to Code, shall not exceed:**

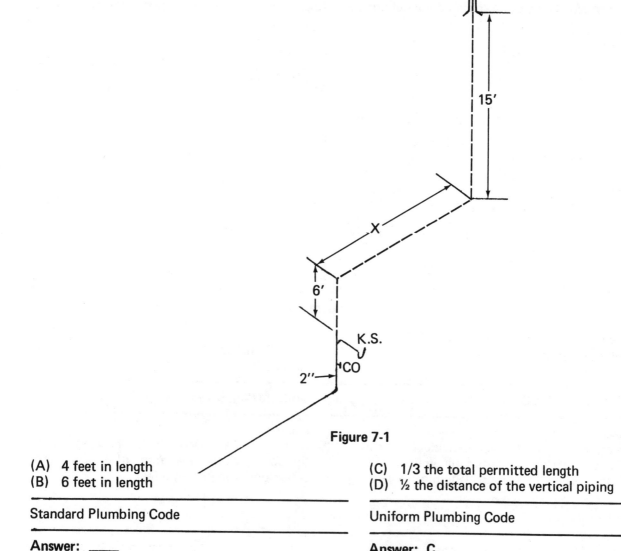

Figure 7-1

(A) 4 feet in length
(B) 6 feet in length

(C) 1/3 the total permitted length
(D) ½ the distance of the vertical piping

Standard Plumbing Code

Uniform Plumbing Code

Answer: _____
Code response: None

Answer: C
Code response: 1/3 of the total permitted length of any vent may be installed in a horizontal position.

Note: The Standard Plumbing Code makes reference to battery-type venting and is not specific as to horizontal dry vents. Review local Code for verification.

7-5 To meet Code requirements, the horizontal portion "X" of the dry vent piping in Figure 7-2 must be installed:

(A) 1 pipe size larger than the Code Vent Table
(B) to drain to the base of the vertical vent stack
(C) no smaller in size than the waste pipe it serves
(D) to drain back by gravity to the drainage pipe it serves

Standard Plumbing Code

Answer: D
Code response: Vent and branch vent piping shall be installed and graded to drip back by gravity to the drainage pipe they serve.

Uniform Plumbing Code

Answer: D
Code response: All vent and branch vent pipes shall be so graded and connected as to drip back by gravity to the drainage pipe they serve.

7-6 According to Code, of the following plumbing piping, the one in which hoarfrost often collects is the:

(A) soil stack
(B) vent stack
(C) waste stack
(D) cold water service

Standard Plumbing Code

Answer: B
Code response: Where there is a possibility of frost or snow closure, the vent extension through a roof shall be protected.

Uniform Plumbing Code

Answer: B
Code response: Where frost or snow closure is likely to occur, each vent extension through the roof shall be protected.

7-7 According to Code, a vent in a building having more than 1 roof level cannot terminate closer than _____ feet to any window opening.

(A) 6
(B) 8
(C) 10
(D) 12

Standard Plumbing Code

Answer: C
Code response: A vent shall not terminate within 10 feet horizontally of any window.

Uniform Plumbing Code

Answer: C
Code response: Each vent shall terminate not less than 10 feet from any window.

7-8 Where frost or snow closure is likely to occur, each vent extension through a roof must be at least 3 inches in diameter. Code specifies that for a 2-inch vent, the change in diameter must be made inside the building at least _____ inches below the roof.

(A) 6
(B) 12
(C) 18
(D) 24

Standard Plumbing Code

Answer: B
Code response: When it is necessary to increase the size of the vent terminal, the change in diameter shall be made no less than 1 foot inside the building.

Uniform Plumbing Code

Answer: B
Code response: The change in diameter of a vent terminal shall be made inside the building at least 1 foot below the roof.

Q 7-9 Vent pipe extensions above a roof must terminate at least ____ inches above it to meet minimum Code requirements.

(A) 4

(B) 6

(C) 8

(D) 10

Standard Plumbing Code

Uniform Plumbing Code

Answer: B

Code response: Vent pipe extensions through a roof shall be terminated at least 6 inches above it.

Answer: B

Code response: Each vent pipe of a stack shall extend through its flashing and shall terminate vertically not less than 6 inches above the roof.

Q 7-10 In high-rise buildings exceeding ____ stories, a relief <u>yoke</u> vent must be installed at differing intervals to meet the Code.

(A) 5

(B) 8

(C) 10

(D) 12

Standard Plumbing Code

Uniform Plumbing Code

Answer: C

Code response: In buildings having more than 10 branch intervals, soil and waste stacks shall be provided with a relief **yoke** vent.

Answer: C

Code response: Each drainage stack which extends 10 or more stories above the building drain or other horizontal drain shall be served by a relief **yoke** vent.

Q 7-11 A 4-inch main vent stack is installed parallel to a 6-inch waste stack in a 32-story building. The Code states that the minimum size of a relief <u>yoke</u> vent that may connect to the main vent stack is ____ inches.

(A) 2

(B) 2 ½

(C) 3

(D) 4

Standard Plumbing Code

Uniform Plumbing Code

Answer: D

Code response: The size of the relief **yoke** vent shall not be smaller than the vent stack to which it connects.

Answer: D

Code response: The size of a relief **yoke** vent shall not be less in diameter than the vent stack to which it is connected.

Q 7-12 According to Code, a relief <u>yoke</u> vent may be installed:

(A) below the fixture branch serving that floor

(B) above the fixture branch serving that floor

(C) at the same level as the fixture branch serving that floor

(D) above the flow line of the fixture branch serving that floor

---

Standard Plumbing Code

Uniform Plumbing Code

Answer: A

Code response: The relief **yoke** vent intersection with the soil or waste stack shall be below the horizontal branch serving the floor.

Answer: A

Code response: The **yoke** vent intersection with the drainage stack shall be placed below the fixture branch serving that floor.

Q 7-13 Where a relief <u>yoke</u> vent is installed in a plumbing system, the Code requires that the relief <u>yoke</u> vent intersection with the drainage pipe be made:

(A) with flexible joints

(B) through a special side-outlet combination fitting

(C) through a wye "Y" branch fitting

(D) through a sanitary tee

Standard Plumbing Code	Uniform Plumbing Code
Answer: C Code response: The relief **yoke** vent shall connect to the soil or waste stack through a wye "Y" fitting.	**Answer: C** Code response: The **yoke** vent intersection with the drainage stack shall be by means of a wye "Y" branch fitting.

Q 7-14 When a relief <u>yoke</u> vent is required by Code in a high-rise building, the relief <u>yoke</u> vent intersection with the main vent pipe must be made with a:

(A) wye "Y" branch, single or double

(B) wye "Y" branch, inverted, single or double

(C) wye "Y" branch, upright, single or double

(D) vent branch, single or double

Standard Plumbing Code	Uniform Plumbing Code
Answer: B Code response: The relief **yoke** vent shall connect to the vent stack through a wye "Y" fitting.	**Answer: B** Code response: The relief **yoke** vent intersection with a vent stack shall be by means of a wye "Y" branch fitting.

Note: The wye "Y" branch, inverted fitting with its opening pointing downward is the correct wye "Y" fitting to use for the intersection with a relief **yoke** vent.

Q 7-15 A 4-inch main vent stack is installed in a 14-story apartment building. The main vent stack does not carry the maximum fixture unit loading. The Code requires the main vent:

(A) to extend full size through the roof

(B) to be combined with vents having lesser fixture unit loadings and then extend through the building roof

(C) to be connected to a stack vent no less than 4 inches in diameter

(D) to be reduced in size for maximum fixture unit loading connected thereto

Standard Plumbing Code	Uniform Plumbing Code
Answer: A Code response: All vent pipes shall extend undiminished in size above the roof.	**Answer: A** Code response: All vent pipes shall extend undiminished in size above the roof.

Q 7-16 The Code permits 2 lavatories side-by-side to be served by a common vent, provided that the fixtures waste separately into:

(A) an approved double fitting

(B) individual sanitary tees

(C) a special hi-low tap fitting

(D) a double straight tee

Standard Plumbing Code	Uniform Plumbing Code
Answer: A Code response: Back-to-back fixtures shall enter the drain (waste pipe) through a sanitary cross, and a common vent shall be provided for each 2 fixtures.	**Answer: A** Code response: 2 fixtures may be served by a common vent pipe when each fixture wastes separately into an approved double fitting.

Q 7-17 According to Code, in lounges and restaurants, traps serving sinks which are part of the equipment for bars, soda fountains and counters:

(A) shall be vented

(B) need not be vented

(C) need not have a vent within 5 feet of the sink

(D) shall be wet-vented

Standard Plumbing Code

Uniform Plumbing Code

Answer: B
Code response: Traps serving sinks which are part of the equipment of bars, soda fountains, and counters need not be vented.

Answer: B
Code response: Traps serving sinks which are part of the equipment of bars, soda fountains and counters need not be vented.

Q 7-18 A vent pipe sometimes has to be run horizontally to avoid a pass-through window installed above a kitchen sink. This may be done, according to Code, if the horizontal vent pipe is at least ____ inches above the flood-level rim of the sink.

(A) 2

(B) 4

(C) 6

(D) 8

Standard Plumbing Code

Uniform Plumbing Code

Answer: C
Code response: The vent pipe shall rise vertically, to a point at least 6 inches above the flood-level rim of the fixture it is venting, before offsetting horizontally.

Answer: C
Code response: Each vent shall rise vertically to a point not less than 6 inches above the flood-level rim of the fixture served, before offsetting horizontally.

Q 7-19 A 2-inch vent pipe is installed to provide air circulation for a chemical waste system. According to Code, the vent pipe:

(A) may connect to the nearest vent pipe of the same size

(B) must connect to the nearest vent pipe, provided it is at least one pipe size larger

(C) must be vertical throughout

(D) must be installed independently through the building roof

Standard Plumbing Code

Uniform Plumbing Code

Answer: D
Code response: Vent piping on acid waste systems shall not be connected to vents of a conventional plumbing system.

Answer: D
Code response: No chemical vent shall intersect vents for other services.

7-20 Only 1 of the following materials is acceptable by Code for venting a chemical waste system. It is:

(A) type-L copper pipe
(B) PVC schedule 40 pipe

(C) galvanized wrought-iron pipe
(D) borosilicate glass

Standard Plumbing Code

Uniform Plumbing Code

Answer: D
Code response: Vent pipes for an acid waste system shall be borosilicate glass.

Answer: D
Code response: Ventilating pipe for corrosive chemicals shall be acid-resistant glass.

Note: There are other types of approved materials for vent piping used for corrosive waste systems, but "D" is the only acceptable one listed here. Check local Code requirements.

7-21 The Code prohibits the installation of 1 of the following plumbing fixtures on a wet vent:

(A) service sinks
(B) lavatories

(C) bathtubs
(D) showers

Standard Plumbing Code

Uniform Plumbing Code

Answer: A
Code response: The waste pipe from 1 or 2 lavatories may be used as a wet vent for 1 or 2 bathtubs or showers.

Answer: A
Code response: Wet venting is limited to vertical drainage piping receiving the discharge from the trap arm of 1 or 2 fixture unit fixtures.

Note: Fixtures installed on a wet vent pipe must be classified as having low fixture unit load values. This generally eliminates fixtures with more than 2 FU's and restricts other fixtures discharging certain types of waste. **Review the Code carefully with regard to wet vents.**

7-22 According to Code, a wet vent cannot be smaller than _____ inches in diameter.

(A) 1¼
(B) 1½

(C) 2
(D) 2½

Standard Plumbing Code

Uniform Plumbing Code

Answer: C
Code response: The wet vent in no case shall be less than 2 inches in diameter.

Answer: C
Code response: But in no case shall a vertical wet vent be smaller than 2 inches.

Q 7-23 Sumps receiving waste from plumbing fixtures, according to Code, shall be vented with a minimum size vent of _____ inches.

(A) 1¼
(B) 1½

(C) 2
(D) 3

Standard Plumbing Code

Uniform Plumbing Code

Answer. B
Code response: Sump vents shall in no case be sized less than 1½ inches in diameter.

Answer: B
Code response: Sump vents shall not be less than 1½ inches in diameter.

Q 7-24 Sumps and receiving tanks must be provided with a local vent, according to Code. The vent pipe:

(A) has to be installed with a minimum of 1 offset
(B) has to be manufactured of cast-iron material only

(C) must extend separately through the roof
(D) may connect to the nearest vent

Standard Plumbing Code

Uniform Plumbing Code

Answer: C
Code response: Vents from a pneumatic ejector (sump) shall be extended separately to the open air.

Answer: C
Code response: Sumps and receiving tank tops shall be provided with a vent pipe which shall extend separately through the roof.

Q 7-25 According to Code, vent piping installed underground shall not be:

(A) DWV copper
(B) PVC schedule 40

(C) cast-iron no-hub
(D) galvanized wrought iron

Standard Plumbing Code

Uniform Plumbing Code

Answer: D
Code response: None

Answer: D
Code response: No galvanized wrought iron shall be used underground.

Note: The Standard Plumbing Code omits galvanized wrought-iron piping from its approved list of underground vent piping materials. The obvious conclusion, then, is that galvanized wrought-iron piping is not acceptable. The correct answer is "D."

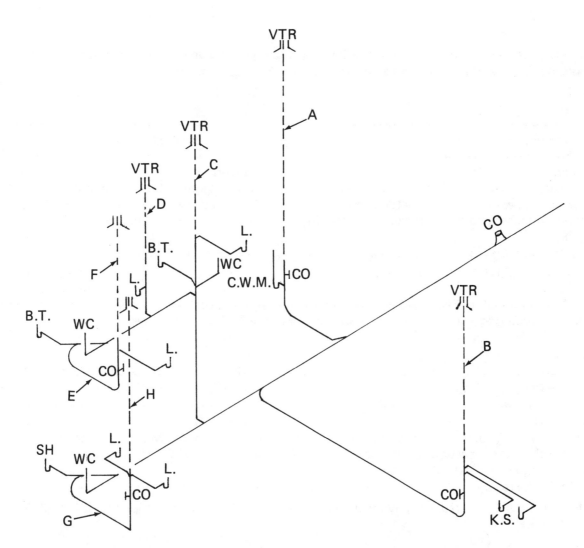

Figure 7-3 shows an isometric drawing for a 3-bath, 2-level residence. The next 8 questions relate to the **minimum** pipe sizes of each listed vent, as required by Code. The 2 Codes do not agree on the sizing requirements for these vents. Be sure to review carefully this section of your local Code. Local codes address these questions directly.

Figure 7-3

7-26 Vent pipe "A" in Figure 7-3 serves a 2-inch waste pipe for a clothes washing machine. According to Code, the minimum size for vent pipe "A" is _____ inches.

(A) 1¼

(B) 1½

(C) 2

(D) 2½

Standard Plumbing Code

Answer: A

Code response: 12 fixture units may connect to a 2-inch vertical waste pipe. Diameter of vent pipe may be 1¼ inches and shall not exceed 30 feet in length.

Uniform Plumbing Code

Answer: B

Code response: 24 fixture units may connect to a 2-inch waste pipe. Diameter of vent pipe shall be 1½ inches and shall not exceed 60 feet in length.

Note: There are 2 different correct answers, depending on which Code is used.

7-27 Vent pipe "B" in Figure 7-3 serves a 2-inch waste pipe for a domestic kitchen sink with dishwasher. According to Code, the minimum size for vent pipe "B" is _____ inches.

(A) 2½

(B) 2

(C) 1½

(D) 1¼

Standard Plumbing Code	Uniform Plumbing Code
Answer: D	**Answer: C**
Code response: 12 fixture units may connect to a 2-inch vertical waste pipe. Diameter of vent pipe may be 1¼ inches and shall not exceed 30 feet in length.	Code response: 24 fixture units may connect to a 2-inch waste pipe. Diameter of vent pipe shall be 1½ inches and shall not exceed 60 feet in length.

Note: There are 2 correct answers, depending on which Code is used.

7-28 Vent pipe "C" in Figure 7-3 is a continuous pipe from the first- and second-floor soil and waste stack. According to Code, the minimum size for vent pipe "C" is _____ inches.

(A) 2

(B) 2½

(C) 3

(D) 4

Standard Plumbing Code	Uniform Plumbing Code
Answer: C	**Answer: C**
Code response: The soil stack shall run undiminished to its connection to the stack vent. Main vent pipes shall extend undiminished in size above the roof.	Code response: Main vent pipes shall extend undiminished in size above the roof.

Note: Each building is required, by most local Codes, to have 1 main vent pipe that is 3 or 4 inches in diameter.

7-29 Vent pipe "D" in Figure 7-3 serves a 2-inch waste pipe for a lavatory. According to Code, the minimum size for vent pipe "D" is _____ inch(es).

(A) 1

(B) 1¼

(C) 1½

(D) 2

Standard Plumbing Code	Uniform Plumbing Code
Answer: B	**Answer: B**
Code response: The diameter of an individual vent shall in no case be less than 1¼ inches.	Code response: The diameter of an individual vent shall not be less than 1¼ inches.

7-30 The horizontal wet vent pipe "E" in Figure 7-3 serves a single bathroom group. According to Code, the minimum size for wet vent pipe "E" is _____ inches.

(A) 1¼
(B) 1½

(C) 2
(D) 2½

Standard Plumbing Code

Uniform Plumbing Code

Answer: B
Code response: A single bathroom group may be served by a 1½-inch wet vent, provided no more than 1 fixture unit is drained into it.

Answer: C
Code response: Wet venting is limited to not more than 4 fixtures on the same floor level. Each wet-vented section shall in no case be less than 2 inches in diameter.

Note: Again, we see a difference of opinion in 2 major Codes. Most local Codes require a 2-inch minimum size wet or dry vent to serve a water closet. Review local Code carefully.

7-31 Vent pipe "F" in Figure 7-3 serves a single bathroom group. According to Code, the minimum size of vent pipe "F" must be _____ inches.

(A) 1¼
(B) 1½

(C) 2
(D) 2½

Standard Plumbing Code

Uniform Plumbing Code

Answer: B
Code response: The vent extension for wet venting a top-floor single bathroom group may be 1½ inches in diameter.

Answer: C
Code response: No water closet shall be vented with a vent pipe less than 2 inches in diameter.

Note: We have two correct answers, depending on which Code is used. Review local Code carefully.

7-32 The horizontal wet vent pipe "G" in Figure 7-3 also serves a single bathroom group. According to Code, the minimum size of wet vent pipe "G" must be _____ inches.

(A) 1¼
(B) 1½

(C) 2
(D) 2½

Standard Plumbing Code

Uniform Plumbing Code

Answer: C
Code response: Up to 4 fixture units may drain into a 2-inch-diameter wet vent.

Answer: C
Code response: Horizontal wet vents shall in no case be less than 2 inches in diameter.

7-33 Vent pipe "H" in Figure 7-3 serves a single bathroom group. According to Code, the minimum size of vent pipe "H" must be _____ inches.

(A) 1¼
(B) 1½

(C) 2
(D) 2½

Standard Plumbing Code

Uniform Plumbing Code

Answer: B
Code response: The vent extension from a single bathroom group may be 1½ inches in diameter.

Answer: C
Code response: No water closet shall be vented with a vent pipe less than 2 inches in diameter.

Note: We have two correct answers, depending on which Code is used. Review local Code carefully.

7-34 Subsoil drainage systems receiving waste from plumbing fixtures, according to Code, must:

(A) have 1 minimum size vent of 2 inches

(B) be sized and vented in accordance with gravity system requirements

(C) be sized and vented at least 1 pipe size larger than gravity system requirements

(D) be considered as a separate system and have 1 minimum size vent of 3 inches.

Standard Plumbing Code

Uniform Plumbing Code

Answer: B
Code response: The drainage and venting system for piping below the sewer level shall be installed in a manner similar to that of the gravity system.

Answer: B
Code response: The drainage and venting systems in connection with fixture sumps, shall be installed under the same requirements as provided for gravity systems.

7-35 According to Code, a _____ is a vent which receives the discharge from waste, except waste from water closets.

(A) wet vent

(B) common vent

(C) loop vent

(D) circuit vent

Standard Plumbing Code

Uniform Plumbing Code

Answer: A
Code response: A wet vent is a vent pipe that receives the discharge from wastes other than water closets.

Answer: A
Code response: A wet vent is vent which also serves as a drain. No 6-unit traps or water closet are permitted to discharge into a wet vent.

7-36 Assume that a water closet is back-vented by a 2-inch-diameter pipe. The vent pipe is connected to a 4-inch by 2-inch sanitary tee. It is graded to drain back to the soil pipe by gravity. According to Code, the isometric drawing in Figure 7-4 is:

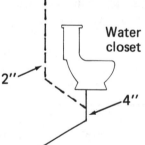

Water closet

2"

4"

Figure 7-4

(A) incorrect. The vent, as illustrated, should be a wet vent

(B) incorrect. First-floor side inlet closet bends are not permitted

(C) incorrect. Horizontal vent connections are prohibited by Code

(D) correct. The installation, as illustrated, complies with Code

Standard Plumbing Code

Uniform Plumbing Code

Answer: D
Code response: Side inlet closet bends are permitted only in cases where the fixture (water closet) connected thereto is vented.

Answer: D
Code response: The vent pipe opening from a soil or waste pipe, **except for water closets and similar fixtures,** shall not be below the weir of the trap.

7-37 Assume that a waste pipe serving a single lavatory is 1½ inches in diameter. The distance between each floor is approximately 10 feet. The plumber installed a 1¼-inch vent through the building roof. According to Code, the isometric drawing in Figure 7-5 is:

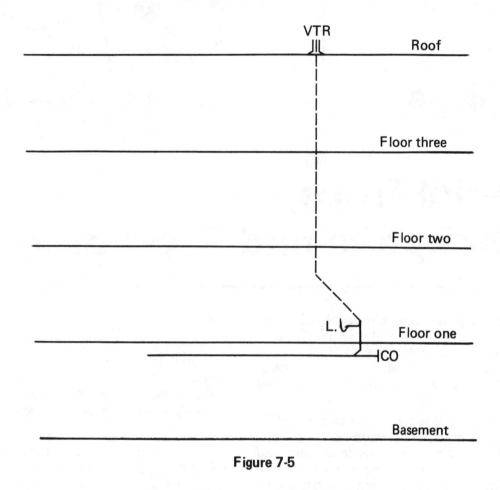

Figure 7-5

(A) correct. The installation, as illustrated, complies with Code

(B) incorrect. The vent, as illustrated, should be vertical throughout

(C) incorrect. The vent pipe should be the same size as the waste pipe

(D) incorrect. The waste and vent pipes should be 1 pipe size larger than illustrated

Standard Plumbing Code

Answer: A
Code response: A 1¼-inch vent pipe with a maximum length of 50 feet may serve up to 8 fixture units.

Uniform Plumbing Code

Answer: A
Code response: 1¼-inch vent pipe with a maximum length of 45 feet may serve up to 8 fixture units.

Chapter 8

Special Traps, Interceptors and Separators

Many substances are considered objectionable waste, or waste that is harmful to the building drainage system, the public sewer, or the sewage treatment plant. These substances include grease, flammables, oils, sand, plaster, lint, hair and ground glass.

Objectionable waste must be intercepted or separated from other liquid waste before it enters the drainage system. The terms *interceptor and separator* mean the same thing and are used to designate the part of the plumbing system that does the separating or intercepting. There are many types of separators, so the type of material separated, the location, or some word indicating the intended use will be added to further identify the item.

Neither the *Standard Plumbing Code* nor the *Uniform Plumbing Code* sets standards, sizes, types, or locations for interceptors. However, both codes emphasize the need to keep certain waste substances out of the drainage system.

Your code will describe exactly the material each interceptor is made of, the size, the type, the installation method, the location, and when interceptors must be used. Your *local code* is the authority here, as always, both when taking the exam and when doing the work. But most of the answers in this chapter will also apply in your community.

8-1 Code permits an oil interceptor, which collects liquid waste from a garage floor, to discharge its waste (after treatment) directly into a:

(A)   catch basin

(B)   building sewer

(C)   building storm sewer

(D)   properly sized dry well

Standard Plumbing Code

Uniform Plumbing Code

Answer: B
Code response: An oil interceptor shall be installed in the drainage system to retain oils so they cannot be admitted into the drainage system.

Answer: B
Code response: Oil interceptors shall be provided when necessary to retain and prevent oils from entering the private or public sewer system.

8-2 Interceptors, according to Code, must be:

(A)   designed so they will not become air-bound

(B)   made of concrete to prevent corrosion

(C)   designed for easy use of ladders to service interceptors

(D)   made with an approved flow control device

Standard Plumbing Code

Uniform Plumbing Code

Answer: A
Code response: Interceptors shall be designed so that they will not become air-bound.

Answer: A
Code response: Interceptors shall be designed so that they will not become air-bound.

8-3 The only one of the following business establishments in which, by Code, an approved oil interceptor is required is:

(A)   public storage garages where floor drainage is to be provided

(B)   public parking decks where floor drainage is to be provided

(C)   manufacturing and assembly plants

(D)   machine shops

Standard Plumbing Code

Uniform Plumbing Code

Answer: A
Code response: An oil separator shall be installed when, in the opinion of the Plumbing Official, a hazard exists and oils can be introduced into the drainage system.

Answer: A
Code response: Oil and sand interceptors shall be provided when, in the judgment of the Administrative Authority, they are necessary for the proper handling of liquid waste or other ingredients harmful to the building drainage system.

Note: According to most local Codes, the only business listed which requires an oil interceptor is the public storage garage. The correct answer is "A."

Q 8-4 When the Code requires that a floor drain be provided in a building where vehicles are repaired, the floor drain must have a minimum _____-inch outlet.

(A) 2
(B) 2½

(C) 3
(D) 4

Standard Plumbing Code	Uniform Plumbing Code
Answer: D Code response: No provision addressing types of floor drains. Local Plumbing Official has jurisdiction in sizing and location.	**Answer: D** Code response: No provision addressing types of floor drains. Local Administrative Authority has jurisdiction in sizing and location.

Note: Most local Plumbing Codes require that floor drains, when installed in a vehicle-repair building, have a minimum 4-inch waste outlet. Answer "D" is correct.

Q 8-5 Where gasoline, oil and sand interceptors are required by Code, the pipe invert of the drain inlet to the interceptor basin must be located not less than _____ inch(es) above the waterline.

(A) ½
(B) 1

(C) 1½
(D) 2

Standard Plumbing Code	Uniform Plumbing Code
Answer: B Code response: No provision addressing design of gasoline, oil and sand interceptors. Local Plumbing Official is responsible for its design.	**Answer: B** Code response: No provision addressing design of gasoline, oil and sand interceptors. Local Administrative Authority is responsible for its design.

Note: As a rule, local Plumbing Codes require that the drain pipe inlet to the interceptor basin be a minimum of 1 inch above the waterline. Answer "B" is correct.

Q 8-6 According to Code, the only one of the following that <u>cannot</u> discharge its waste through a grease interceptor is a:

(A) garbage-can washer
(B) drain from a hand sink

(C) food-waste disposal
(D) drain from a drink cooler

Standard Plumbing Code	Uniform Plumbing Code
Answer: C Code response: Where food-waste grinders are installed, the waste from those units shall not pass through a grease interceptor.	**Answer: C** Code response: No food-waste disposal unit shall discharge into any grease interceptor.

The next 8 questions are based on drawings or sectional views of various types of special traps, interceptors and separators used in plumbing drainage systems. Examine each question and drawing carefully before selecting your answer.

Q 8-7 The interceptor shown in Figure 8-1 is required by Code to prevent _____ from entering into the building drainage system.

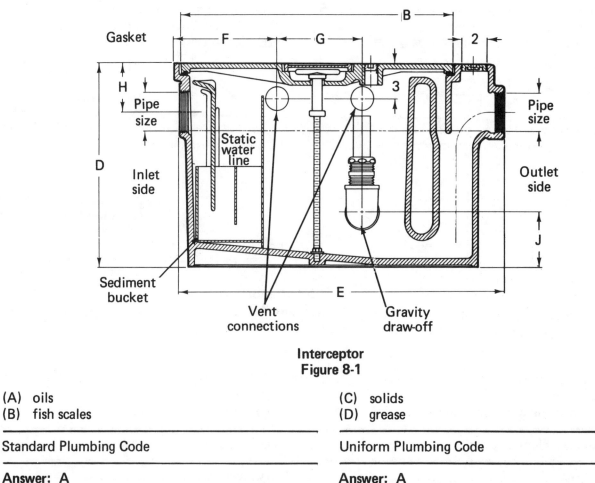

Interceptor
Figure 8-1

(A) oils
(B) fish scales

(C) solids
(D) grease

Standard Plumbing Code	Uniform Plumbing Code

Answer: A
Code response: No provision addressing design of interceptor. Local Plumbing Official has sole jurisdiction.

Answer: A
Code response: No provision addressing design of interceptor. Local Administrative Authority has sole jurisdiction.

Note: Figure 8-1 illustrates an approved oil interceptor, manufactured for use in certain areas of the country. This interceptor is accepted by many local Codes for installation in drain lines of garages and service stations. Answer "A" is correct.

Q 8-8 The interceptor shown in Figure 8-2 is required by Code to be installed in a drain line to prevent the introduction of _____ into the building drainage system.

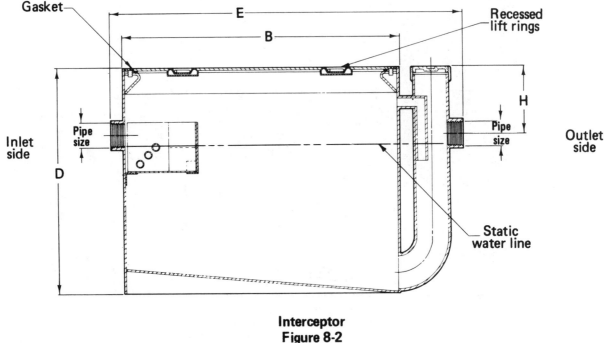

Interceptor
Figure 8-2

(A) broken glass
(B) alkaline waste substances

(C) grease
(D) sand

Standard Plumbing Code

Answer: C
Code response: No provision addressing design of interceptors. Local Plumbing Official has sole jurisdiction.

Uniform Plumbing Code

Answer: C
Code response: No provision addressing design of interceptors. Local Administrative Authority has sole jurisdiction.

Note: Figure 8-2 illustrates an approved grease interceptor, manufactured for use in many areas of the country. It is accepted by many local Codes for installation in drain lines of such business establishments as restaurants and hotel kitchens. Answer "C" is correct.

Q **8-9** The interceptor shown in Figure 8-3 is usually required, by Code, to be installed in a drain line to prevent _____ from entering the building drainage system.

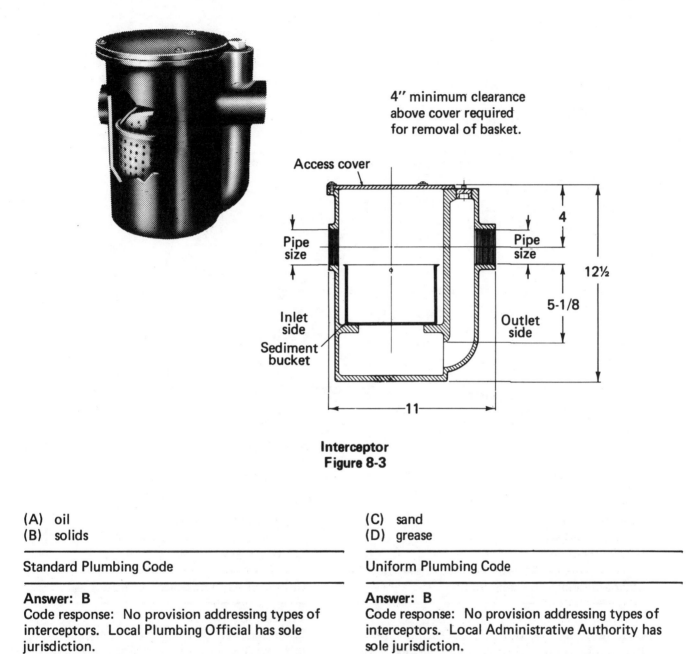

4" minimum clearance above cover required for removal of basket.

Access cover

Pipe size

Pipe size

4

12½

5-1/8

Inlet side

Outlet side

Sediment bucket

11

Interceptor
Figure 8-3

(A) oil
(B) solids

(C) sand
(D) grease

Standard Plumbing Code

Uniform Plumbing Code

Answer: B
Code response: No provision addressing types of interceptors. Local Plumbing Official has sole jurisdiction.

Answer: B
Code response: No provision addressing types of interceptors. Local Administrative Authority has sole jurisdiction.

Note: Figure 8-3 shows an approved solids interceptor with a removable perforated sediment bucket. It's almost certain that local Code will require an interceptor to be installed in drain lines for slaughter houses, meat markets and fish markets. Answer "B" is correct.

Q **8-10** The interceptor shown in Figure 8-4 is usually installed in _____ waste lines when required or permitted by Code.

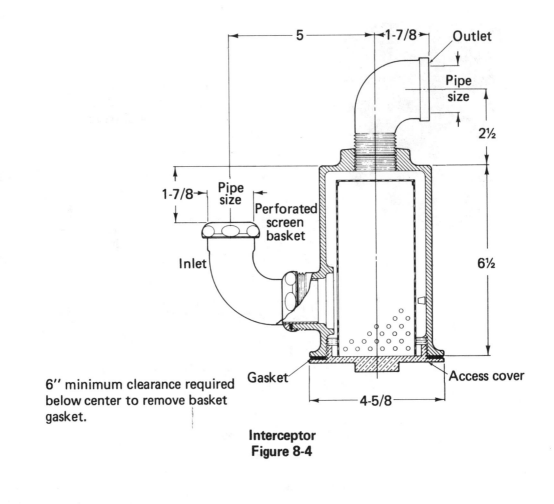

6" minimum clearance required
below center to remove basket
gasket.

Interceptor
Figure 8-4

(A) laboratory sink
(B) developing sink

(C) orthopedic sink
(D) barber shop sink

Standard Plumbing Code	Uniform Plumbing Code
Answer: D	**Answer: D**
Code response: No provision addressing types of interceptors. Local Plumbing Official has sole jurisdiction.	Code response: No provision addressing types of interceptors. Local Administrative Authority has sole jurisdiction.

Note: Figure 8-4 shows a hair interceptor with a removable screen basket. Local Codes generally require it where large quantities of hair may be introduced into the drainage system. This includes barber shop sinks, beauty salon sinks, or fixtures used for bathing animals. Answer "D" is correct.

8-11 The special interceptor floor drain in Figure 8-5 would, according to Code, be most likely installed in the waste line or lines of:

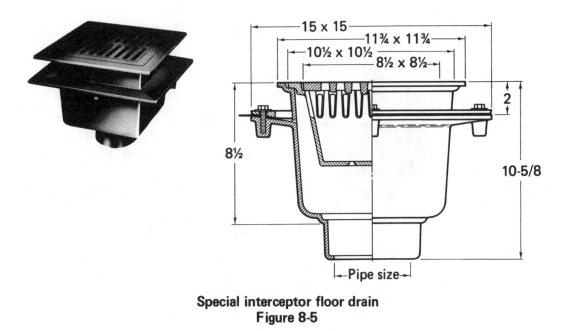

Special interceptor floor drain
Figure 8-5

(A) a building where vehicles are repaired
(B) a dog kennel

(C) a restaurant
(D) a public toilet room

Standard Plumbing Code

Answer: A
Code response: No provision addressing special interceptor floor drain. Local Plumbing Official has sole jurisdiction in determining its use.

Uniform Plumbing Code

Answer: A
Code response: No provision addressing special interceptor floor drain. Local Administrative Authority has sole jurisdiction in determining its use.

Note: Figure 8-5 shows a floor drain with sediment bucket. Local Codes usually require it where floor drainage is provided in garages, service stations or where vehicles are repaired. Answer "A" is correct.

8-12 The special fixture shown in Figure 8-6, according to Code, would be installed to receive:

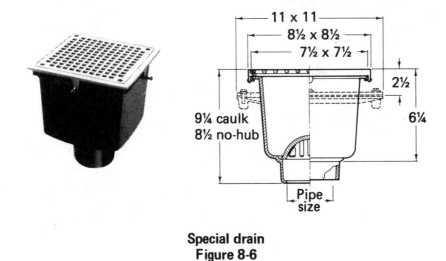

**Special drain
Figure 8-6**

(A) kitchen floor drainage

(B) deck water around commercial swimming pools

(C) floor drainage from commercial parking garages

(D) waste drainage from a beverage cooler

Standard Plumbing Code	Uniform Plumbing Code
Answer: D Code response: No provision addressing special fixture. Local Plumbing Official has sole jurisdiction in approving its use.	**Answer: D** Code response: No provision addressing special fixture. Local Administrative Authority has sole jurisdiction in approving its use.

Note: Figure 8-6 illustrates a square floor sink approved by local Codes for use in restaurants or similar establishments to receive waste from special fixtures and equipment by indirect means. Answer "D" is correct.

Q 8-13 The special trap shown in Figure 8-7 would be installed, according to Code, on a:

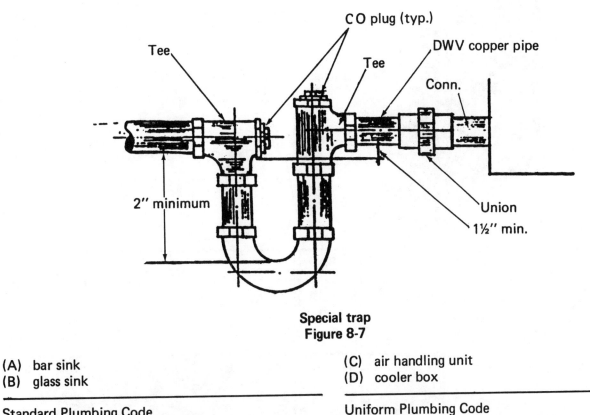

**Special trap
Figure 8-7**

(A) bar sink

(B) glass sink

(C) air handling unit

(D) cooler box

Standard Plumbing Code

Answer: C
Code response: No provision addressing air-handling unit traps. Local Plumbing Official has sole jurisdiction in approving its use.

Uniform Plumbing Code

Answer: C
Code response: No provision addressing air-handling unit traps. Local Administrative Authority has sole jurisdiction in approving its use.

Note: Local Codes almost always require that a trap be installed on air-handling units. Figure 8-7 shows a job-prefabricated air-handling unit trap, Answer "C" is correct.

Q 8-14 The special interceptor trap shown in Figure 8-8 would, according to Code, be a required installation in the waste line of a:

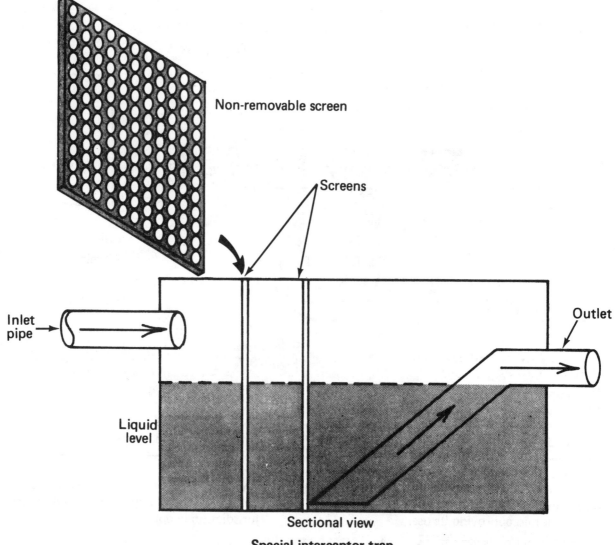

Non-removable screen

Screens

Inlet pipe

Outlet

Liquid level

Sectional view
Special interceptor trap
Figure 8-8

(A) bottle plant
(B) commercial laundry

(C) dry cleaning establishment
(D) veterinary clinic

Standard Plumbing Code

Answer: B
Code response: Commercial laundries shall have installed an interceptor to collect solids to prevent their passing into the drainage system.

Note: Local Code will probably require that a lint interceptor be installed in the waste lines of commercial laundries. This prevents strings, rags and buttons, etc. from entering the drainage system. Figure 8-8 shows a job-prefabricated lint interceptor, usually not available from manufacturers. Answer "B" is correct.

Uniform Plumbing Code

Answer: B
Code response: No provision addressing interceptors for commercial laundries. Local Administrative Authority has sole jurisdiction in approving its use.

Q 8-15 According to Code, interceptors shall be installed:

(A) by a licensed plumbing contractor
(B) by a licensed septic tank contractor

(C) where they are accessible
(D) next to a 3-compartment pot sink

Standard Plumbing Code

Uniform Plumbing Code

Answer: C
Code response: Each interceptor shall be installed so the top is readily accessible for servicing and maintaining the interceptor in working and operating condition.

Answer: C
Code response: Each interceptor cover shall be readily accessible for servicing and maintaining the interceptor in working and operating condition.

Q 8-16 According to Code, wastes exceeding ____ should not discharge into a grease interceptor.

(A) 130 degrees F.
(B) 140 degrees F.

(C) 150 degrees F.
(D) 160 degrees F.

Standard Plumbing Code

Uniform Plumbing Code

Answer: B
Code response: Water above 140 degrees Fahrenheit shall not discharge directly into any part of a drainage system.

Answer: B
Code response: Wastes in excess of 140 degrees Fahrenheit shall not discharge into a grease trap.

Q 8-17 Corrosive liquids must pass through an approved dilution or neutralizing tank before discharge into the regular sanitary system. According to Code, only a tank that is ____ is acceptable.

(A) copper-lined
(B) stainless steel

(C) earthenware
(D) extra-heavy cast-iron

Standard Plumbing Code

Uniform Plumbing Code

Answer: C
Code response: Acids, corrosive liquids and chemicals must be thoroughly diluted by passing through a properly constructed and acceptable dilution or neutralizing device.

Answer: C
Code response: Chemical or industrial liquid wastes shall be pretreated to render them innocuous prior to discharge into a drainage system.

Note: Dilution or neutralizing tanks are normally constructed of earthenware or glass. Corrosive liquids will not harm these tanks.

Q 8-18 According to Code, the special trap that can be used under certain conditions, if first approved by the local Administrative Authority having jurisdiction, is the:

(A) drum trap
(B) "S" trap

(C) bell trap
(D) partitioned trap

Standard Plumbing Code

Uniform Plumbing Code

Answer: A
Code response: Drum traps shall be limited to special fixtures and are subject to approval by the Plumbing Official.

Answer: A
Code response: Drum traps may be installed only when permitted by the Administrative Authority for special conditions.

8-19 The requirements for installation of grease interceptors, according to Code, do not include:

(A) fast-food restaurants
(B) apartment buildings with up to 100 dwelling units

(C) hospitals
(D) hotels

Standard Plumbing Code

Answer: B
Code response: A grease interceptor shall be installed in the waste line leading from businesses that prepare or serve food on a commercial basis.

Uniform Plumbing Code

Answer: B
Code response: A grease trap is not required for individual dwelling units.

8-20 Manufactured cast-iron grease interceptors may serve a single commercial sink. To insure effectiveness, the wastes entering the grease interceptor must not exceed the manufacturer's rated capacity. For this reason, the Code requires the installation of a:

(A) clarifier
(B) balancing valve

(C) swing check valve
(D) flow control

Standard Plumbing Code

Answer: ____
Code response: No provision addressing flow control for grease interceptor. Local Plumbing Official and the manufacturer have jurisdiction.

Uniform Plumbing Code

Answer: D
Code response: Each plumbing fixture connected to a grease trap shall be provided with an approved flow control.

Note: Manufactured grease interceptors are acceptable in many areas of the country. Where used, a flow control is furnished as standard equipment for the interceptor. This controls the flow rate into the interceptor for proper separation of grease from other liquid wastes. Answer "D" is correct.

8-21 Assume an interceptor has a retention capacity of 40 pounds of grease. The Code would enforce a maximum flow rate in gpm as recommended by the manufacturer. The approved rate of flow would be ____ gallons per minute.

(A) 10
(B) 15

(C) 20
(D) 25

Standard Plumbing Code

Answer: ____
Code response: No provision addressing rate of flow for grease interceptor. Local Plumbing Official and the manufacturer have jurisdiction.

Uniform Plumbing Code

Answer: C
Code response: The approved rate of flow for a grease interceptor having a retention capacity of 40 pounds is 20 gpm.

Note: The Standard Plumbing Code does not address the specifics for grease interceptors. Use the manufacturer's recommendations. Answer "C" is correct.

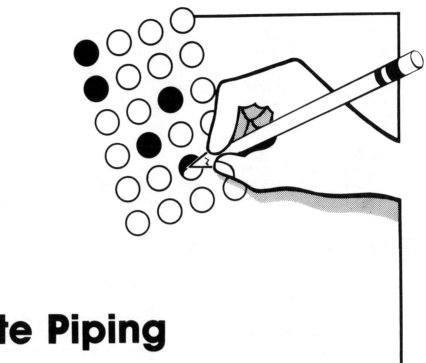

Chapter 9

Indirect and Special Waste Piping

It's not practical to use conventional waste and vent piping for many of the fixtures, appliances and devices not regularly classed as plumbing fixtures. The code allows certain special fixtures, appliances and devices with drips or drains to be *indirectly* connected to a building drainage system.

These appliances and fixtures include refrigerators, ice boxes, bar sinks, cooling or refrigerating coils, laundry washers, extractors, steam tables, egg boilers, coffee urns, stills, sterilizers, water stations, water lifts, expansion tanks, cooling jackets, drips or overflow pans, air conditioning condensate drains, drains from overflows, and relief vents from the water supply system.

Indirect drainage must still keep sewage from backing up into the fixture and prevent contamination of the contents in the case of a stoppage in the sanitary drainage system. Overflow and relief pipes on the water supply system and relief pipes on expansion tanks and cooling jackets must always be indirectly connected to the sanitary drainage system. This prevents the possibility of a cross-connection which could contaminate the potable water supply system.

The code also requires and enforces a positive separation (air gap) by indirect means between the waste outlet and the building drainage system of hospital, food storage and food preparation equipment. This unique method of piping is used most frequently in restaurants, food preparation and packaging establishments, and hospitals.

Special waste and vent piping includes acid, chemical or industrial wastes, vertical or horizontal wet venting, and combination waste and vent systems. In most cases, prior approval from the plumbing authority is required before actual installation of these systems.

The *Standard Plumbing Code* and the *Uniform Plumbing Code* don't spell out all the specifics necessary for the installation of indirect waste and special waste piping. This leaves much of the interpretation about material types, sizes, lengths of runs and installation methods in the hands of the plumbing inspector and the government agency that adopted your local code. Review this section thoroughly in your local code.

9-1 According to Code, when floor drainage is required for a cold storage room used for storing or holding food or drink, the drain must be connected to the sanitary drainage system through a:

(A)   direct connection

(B)   deep seal trap with trap primer

(C)   floor drain with basket

(D)   indirect waste

Standard Plumbing Code

Uniform Plumbing Code

Answer: D

Code response: Floor drainage used for storing or holding food shall discharge into the building drainage system through an indirect waste.

Answer: D

Code response: Cold storage rooms used for the storage or holding of food or drink shall be drained by means of indirect waste pipes.

9-2 According to Code, the maximum height above the trap for a clothes washer standpipe receptor is _____ inches.

(A)   18

(B)   26

(C)   30

(D)   48

Standard Plumbing Code

Uniform Plumbing Code

Answer: D

Code response: The vertical distance of a clothes washer standpipe receptor shall not exceed 48 inches above its trap.

Answer: C

Code response: No standpipe receptor for any clothes washer shall extend more than 30 inches above its trap.

Note: There are two correct answers, depending on which Code is used. Always review local Code for specifics on standpipe receptors.

9-3 According to Code, the minimum height above the trap for a domestic clothes washer standpipe receptor is _____ inches.

(A)   12

(B)   18

(C)   20

(D)   24

Standard Plumbing Code

Uniform Plumbing Code

Answer: _____

Code response: None

Answer: B

Code response: No standpipe receptor for any clothes washer shall be less than 18 inches above its trap.

Note: The Standard Plumbing Code and many local Codes do not address the minimum height above the trap for a clothes washer standpipe receptor. But domestic clothes washer standpipe receptors are normally manufactured to a standard height. Review your local Code for requirements in your area.

Q 9-4 To meet Code, the trap for a clothes washer must be roughed-in not less than _____ inches above the floor.

(A) 4
(B) 6

(C) 8
(D) 10

___

Standard Plumbing Code

Uniform Plumbing Code

Answer: _____
Code response: None

Answer: B
Code response: No trap for any clothes washer standpipe receptor shall be roughed-in less than 6 inches above the floor.

Note: The Standard Plumbing Code and many local Codes do not make reference to the above-floor height required to rough-in a waste outlet for a clothes washer. All Codes prohibit the installation of a trap below a floor. Review your local Code for requirements in your area.

Q 9-5 No indirect waste receptor, according to Code, can be installed in any:

(A) bedroom
(B) storeroom

(C) dining room
(D) family room

___

Standard Plumbing Code

Uniform Plumbing Code

Answer: B
Code response: No waste receptors serving indirect waste pipes shall be installed in any storeroom.

Answer: B
Code response: No indirect waste receptor shall be installed in any storeroom.

Q 9-6 Code requires that indirect waste piping from coffee urns <u>not</u> drain into a waste receptor located in a:

(A) public toilet room
(B) restaurant dining room

(C) lounge where alcohol is served
(D) area where a salad bar is located

___

Standard Plumbing Code

Uniform Plumbing Code

Answer: A
Code response: No waste receptors serving indirect waste pipes shall be installed in any toilet room.

Answer: A
Code response: No indirect waste receptor shall be installed in any toilet room.

Q 9-7 To properly connect a steam pipe to a plumbing drainage system, it is mandated by Code that:

(A) the pipe be not less than 10 feet in length

(B) the diameter of the pipe be a minimum of 1¼ inches

(C) a direct connection be made to a waste stack only

(D) an indirect connection be made

___

Standard Plumbing Code

Uniform Plumbing Code

Answer: D
Code response: A steam pipe or piping shall not connect directly to any part of a drainage system.

Answer: D
Code response: No steam pipe shall be directly connected to any part of a plumbing drainage system.

9-8 Code mandates that receptors receiving the discharge of indirect waste pipes be of a shape and capacity that will:

(A) prevent splashing
(B) serve the intended use

(C) avoid collecting solids
(D) intercept ingredients harmful to the building drainage system

---

Standard Plumbing Code

Uniform Plumbing Code

Answer: A
Code response: Plumbing receptors receiving the discharge of indirect waste pipes shall be of such shape and capacity as to prevent splashing.

Answer: A
Code response: All receptors receiving the discharge of indirect waste pipes shall be of such shape and capacity as to prevent splashing.

9-9 Assume you are installing a waste pipe for a culinary sink. The pipe is a 1½-inch copper pipe. The Code requires that the sink be protected from backflow. In order to pass inspection, you must provide:

(A) a check valve
(B) an air gap

(C) a backwater valve
(D) a vacuum breaker

Standard Plumbing Code

Uniform Plumbing Code

Answer: B
Code response: A culinary sink drain shall be indirectly connected to the drainage system.

Answer: B
Code response: No culinary sink shall have any drain directly connected to any waste pipe.

9-10 According to Code, the drain pipe from a refrigerator must drain into an:

(A) approved dry well
(B) catch basin

(C) open floor sink
(D) service sink

Standard Plumbing Code

Uniform Plumbing Code

Answer: C
Code response: A refrigerator shall discharge indirectly into a water-supplied (floor) sink.

Answer: C
Code response: A refrigerator shall be drained by means of indirect waste pipes into an open floor sink.

9-11 Code mandates that the waste pipe from a domestic dishwashing machine connected to a food-waste disposer unit have a:

(A) soft-seat check valve
(B) approved air-gap fitting

(C) backflow preventer valve
(D) nonrising stem gate valve

Standard Plumbing Code

Uniform Plumbing Code

Answer: B
Code response: Domestic dishwashing machines shall not be directly connected to a drainage system.

Answer: B
Code response: No domestic dishwashing machine shall be directly connected to a food-waste disposer without the use of an approved dishwasher air-gap fitting on the discharge side of the dishwashing machine.

Note: The Standard Plumbing Code does not specifically require the installation of an air-gap fitting, nor does it prohibit its use. However, the Code **does** require an indirect connection for dishwashers, thereby protecting the dishwasher from backflow. Answer "B" is correct.

9-12 A freestanding drinking fountain, according to Code, may connect to the drainage system of a building:

(A) by using an approved air-gap fitting

(B) by an indirect waste method

(C) through an individual P-trap only

(D) provided the waste outlet is no more than 10 feet from a vent

---

Standard Plumbing Code

Uniform Plumbing Code

Answer: B
Code response: Drinking fountains may be installed with indirect wastes.

Answer: B
Code response: Drinking fountains may be installed with indirect wastes.

9-13 When any Code requires that indirect waste pipes be vented, the vent:

(A) must be of noncorrosive material

(B) must not be smaller than the indirect waste pipe it serves

(C) must extend separately up and through the roof

(D) may connect to the nearest vent

Standard Plumbing Code

Uniform Plumbing Code

Answer: C
Code response: Vent pipes on acid and chemical **indirect** waste pipes shall not connect to the conventional plumbing system.

Answer: C
Code response: No vent from indirect waste piping shall combine with any sewer-connected vent but shall extend separately to the outside air.

Note: The Standard Plumbing Code specifically addresses only acid or chemical venting of indirect waste pipes, not the conventional venting. Answer "C" is correct.

9-14 Air conditioning equipment, according to Code, may connect:

(A) directly to a vent pipe

(B) indirectly to a vent pipe

(C) into a sink tailpiece, if first approved by Plumbing Official

(D) by indirect waste pipes to the drainage system

Standard Plumbing Code

Uniform Plumbing Code

Answer: D
Code response: Air conditioning units, when connected to the building drainage system, shall be by indirect means and shall be classified as a plumbing fixture.

Answer: D
Code response: Air conditioning equipment wastes shall be drained by indirect waste pipes into the drainage system.

Q 9-15 When listed air-gap fittings are required by Code for sinks receiving the wastes from dishwashing machines, the plumber must install the flood-level marking:

(A) above the flood level of the sink
(B) above the sink drainboard

(C) A or B above, whichever is higher
(D) facing away from the backsplash

Standard Plumbing Code	Uniform Plumbing Code
Answer: C Code response: Domestic dishwashing machines shall be protected from backflow by providing an air gap in the drain connection on the inlet side of the trap serving the appliance.	**Answer: C** Code response: Listed air gaps shall be installed with the flood-level marking at or above the flood level of the sink or drainboard, whichever is higher.

Note: Look again at the "note" for question 9-11. Compare the wording in 9-11 to the wording in 9-15. For question 9-15, "C" is the correct answer.

Q 9-16 A sterilizer is required in a hospital surgical room. It is used for sterile materials and requires a water and waste connection. According to Code, such waste must:

(A) pass through a cooling tank before connecting to the drainage system
(B) be indirectly connected to the drainage system

(C) be neutralized before it is connected to the drainage system
(D) be connected directly to a separate trap

Standard Plumbing Code	Uniform Plumbing Code
Answer: B Code response: Sterilizers requiring water and waste connections and used for sterile material shall be indirectly connected to the drainage system.	**Answer: B** Code response: Sterilizers requiring water and waste and used for sterile materials shall be indirectly connected to the drainage system.

Q 9-17 A bar sink having a 1¼-inch-diameter drain must be connected indirectly to the drainage system. The bar sink discharge capacity is rated at 7½ gallons per minute. According to Code, the diameter of the indirect waste piping cannot be smaller than _____ inch(es).

(A) 1
(B) 1-1/8

(C) 1¼
(D) 1½

Standard Plumbing Code	Uniform Plumbing Code
Answer: C Code response: The material and size of indirect waste pipes shall be in accordance with the provisions of the other sections of this Code applicable to sanitary drainage piping.	**Answer: C** Code response: Indirect waste pipes need not be larger in diameter than the drain outlet or tailpiece of the fixture.

Note: The Standard Plumbing Code does not specifically address the indirect waste pipe size but refers to the sanitary drainage piping table. The minimum size horizontal fixture drain listed therein is 1¼ inches in diameter. Answer "C" is correct.

Q 9-18 The drain pipe from a water station installed in a restaurant is ¾-inch in diameter. With the exception of piping materials constructed of PVC schedule 40 or DWV copper, the Code-required maximum size indirect piping to use is _____ inch(es).

(A) 3/4

(B) 7/8

(C) 1

(D) 1¼

Standard Plumbing Code

Answer: _____
Code response: The size of indirect waste pipes shall be in accordance with the provisions of other sections of this Code applicable to sanitary drainage systems.

Note: The Standard Plumbing Code does not specifically address the above question. The Plumbing Official or local Code usually provides pertinent information. Answer "A" is correct.

Uniform Plumbing Code

Answer: A
Code response: Indirect waste pipe need be no larger in diameter than the drain outlet it serves.

Q 9-19 Code mandates that no plumbing fixtures served by indirect waste pipes or receiving discharge therefrom be:

(A) located in any corridor

(B) placed in use until tested and proved watertight

(C) located in any heavy traffic area

(D) installed until first approved by local authority

Standard Plumbing Code

Answer: _____
Code response: None

Uniform Plumbing Code

Answer: D
Code response: No plumbing fixtures served by indirect waste pipes or receiving discharge therefrom shall be installed until first approved by the Administrative Authority.

Note: The Standard Plumbing Code does not address this particular question. Plumbing Official or local Code usually provides specifics regarding indirect waste pipes. Answer "D" is correct.

Q 9-20 Code requires that relief vents from the water supply system, when connected to the plumbing drainage system, be protected by:

(A) installing a check valve

(B) an indirect connection

(C) a deep seal trap

(D) installing an atmospheric vacuum breaker

Standard Plumbing Code

Answer: B
Code response: Indirect waste connections shall be provided for relief vents from the water supply system.

Uniform Plumbing Code

Answer: B
Code response: Indirect waste connections shall be provided for relief vents from the water supply system.

Q 9-21 Indirect waste piping must be installed, according to Code,

(A) to drain dry
(B) a minimum of 6 inches above the floor
(C) only where it is accessible
(D) with adequate cleanouts to permit cleaning

Standard Plumbing Code | Uniform Plumbing Code

Answer: D
Code response: Indirect waste piping shall be so installed as to permit ready access for flushing and cleaning.

Answer: D
Code response: Indirect waste pipes shall be provided with cleanouts so as to permit flushing and cleaning.

Q 9-22 Floor sinks installed to receive the discharge from indirect waste pipes shall be, according to Code,

(A) set ¼-inch above the finished floor
(B) as close as possible to the appliances they serve
(C) constructed of pervious materials
(D) readily accessible

Standard Plumbing Code | Uniform Plumbing Code

Answer: D
Code response: Suitable fixtures receiving the discharge of indirect waste pipes shall be accessible.

Answer: D
Code response: Receptors receiving the discharge of indirect waste pipes shall be located where they are readily accessible.

Q 9-23 The vertical piping between any 2 consecutive inlet levels is to be considered a ____, according to Code.

(A) soil pipe
(B) waste pipe
(C) wet-vented pipe
(D) vertical vent

Standard Plumbing Code | Uniform Plumbing Code

Answer: C
Code response: A wet vent is a vent (vertical or horizontal) which receives the discharge from waste other than water closets.

Answer: C
Code response: The vertical piping between any 2 consecutive inlet levels shall be considered a wet-vented section.

Note: Figure 9-1 illustrates the above question. Although the Standard Plumbing Code does not specifically describe the above as a wet vent, it is considered a wet vent as described in other sections of their Code.

Q 9-24 The wet-vented section "X", as illustrated in Figure 9-1, must have a minimum ____-inch diameter, according to Code.

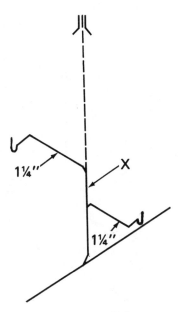

Figure 9-1

(A) 1¼
(B) 1½

(C) 1¾
(D) 2

Standard Plumbing Code

Answer: B
Code response: Two fixtures set on the same floor level but connecting at different levels in the stack, the vertical drain (wet vent section) shall be one pipe size larger than the upper fixture drain.

Uniform Plumbing Code

Answer: B
Code response: Each wet-vented section shall be a minimum of 1 pipe size larger than the required minimum waste pipe of the upper fixture.

Q 9-25 According to Code, the common vent illustrated in Figure 9-2 shall <u>in no case</u> be smaller than _____ inch(es).

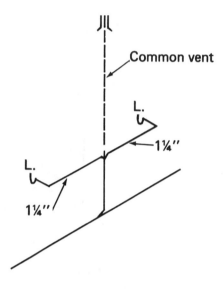

Figure 9-2

(A) 1	(C) 1½
(B) 1¼	(D) 2

Standard Plumbing Code

Answer: B
Code response: The diameter of an individual vent (common vent) shall in no case be smaller than 1¼ inches in diameter.

Uniform Plumbing Code

Answer: B
Code response: Common vent sizing shall in no case be smaller than the minimum vent pipe size required for any fixture served (1¼ inches).

Q 9-26 The isometric drawing in Figure 9-3 is, according to Code:

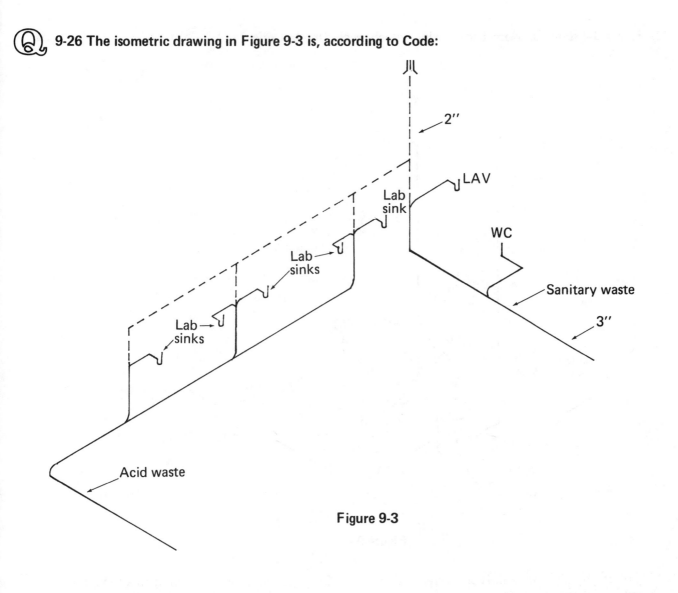

Figure 9-3

(A) incorrect; the water closet should not be wet vented

(B) incorrect; the acid waste loop vent should not connect to a vent pipe serving conventional plumbing fixtures

(C) correct; the systems as illustrated would comply with Code if the vent pipe through the roof were 3 inches

(D) incorrect; only 3 lab sinks are permitted on this type installation

---

Standard Plumbing Code

Answer: B
Code response: Vent piping on acid waste systems shall not be connected to the conventional plumbing system.

Uniform Plumbing Code

Answer: B
Code response: No chemical vent shall intersect vents for other services.

Q 9-27 Figure 9-4 shows an isometric drawing, which, according to Code, is:

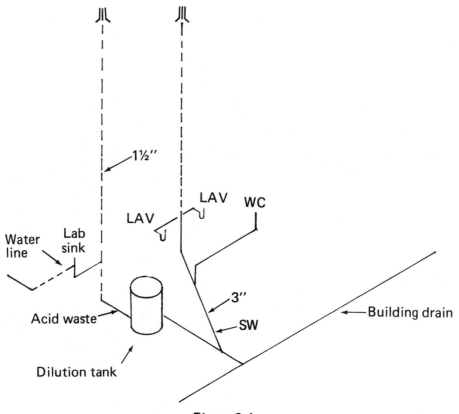

Figure 9-4

(A) correct; the dilution tank is properly located and complies with Code

(B) incorrect; the sanitary waste line should connect directly to the building drain

(C) incorrect; for a system of this type, the acid vent pipe should be no smaller than 3 inches in diameter

(D) incorrect; a dilution tank is not required if a water closet intersects an acid waste line before discharging into the building drainage system

---

Standard Plumbing Code

Answer: A
Code response: In no case shall corrosive liquids or spent acids be discharged into the plumbing drainage system until they have been thoroughly diluted or neutralized by passing through a properly constructed and acceptable dilution or neutralizing device.

Uniform Plumbing Code

Answer: A
Code response: Chemical or industrial liquid wastes shall be pretreated to render them innocuous prior to discharge into the drainage system.

Q 9-28 A horizontal combination waste and vent pipe is to be installed in the kitchen portion of a restaurant. The horizontal main waste pipe is 6 inches. The minimum size of the downstream vent, according to Code, must be no less than _____ inches.

(A) 1½

(B) 2

(C) 2½

(D) 3

Standard Plumbing Code

Answer: D
Code response: The cross-sectional area of the vent shall be not less than ½ of the area of the waste pipe served.

Uniform Plumbing Code

Answer: D
Code response: The minimum area of any vent installed in a combination waste and vent system shall be at least ½ the inside cross-sectional area of the drain pipe served.

Q 9-29 When designing a horizontal waste and vent system, the waste piping, according to Code, must be:

(A) at least 3 inches in diameter

(B) graded a minimum of ¼-inch fall per foot

(C) at least 2 pipe sizes larger than conventional pipe sizes

(D) no less than standard weight pipe

Standard Plumbing Code

Answer: C
Code response: Every waste pipe in this system shall be at least 2 pipe sizes larger than the regular required sizes of a conventional system.

Uniform Plumbing Code

Answer: C
Code response: Each waste pipe in any such system shall be at least 2 pipe sizes larger than the sizes normally required by this Code.

Q 9-30 In designing a horizontal waste and vent system, the Code places a maximum drop between the fixture and its trap. This distance must not exceed _____ inches.

(A) 12

(B) 18

(C) 20

(D) 24

Standard Plumbing Code

Answer: D
Code response: The vertical distance from fixture or drain outlet to trap weir shall not exceed 24 inches.

Uniform Plumbing Code

Answer: D
Code response: The vertical distance between the tailpiece or connection and the outlet of a plumbing fixture and the trap therefor, shall in no case exceed 2 feet.

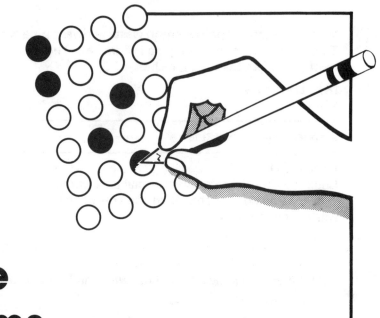

Chapter 10

Private Sewage Disposal Systems

Where public sewers are not available, the most common system for sewage disposal is the septic tank. Some codes still permit disposing of sewage in cesspools for limited, minor or temporary use, but only when the local authority has given advance approval.

A septic tank is a watertight receptacle that receives the wastes from a drainage system through an inlet tee. It's designed to separate solid from liquid wastes. There's usually only about 3/4 of a pound of solids in each 100 gallons of water. Solids are heavier and settle to the bottom of the tank. Liquids, smaller particles and grease rise to the top of the tank.

Anaerobic bacteria, which feed in the absence of air or free oxygen, decompose the solids at the bottom of the tank, changing them into gases and harmless liquids. The gases escape, agitating the tank contents and speeding the action of the bacteria. As new sewage flows through the plumbing system and into the septic tank, the gases are forced up and through the drainage vent pipes and into the atmosphere above the building roof.

The tank must be big enough to hold about as much sewage as can be expected to flow into the tank in 24 hours. This gives the bacteria enough time to fully digest the solids.

The final product of all this is a clear liquid, called *effluent*, which is forced through the outlet tee and into the drainfield. This is a system of open-joint or perforated pipe installed on a bed of washed rock, gravel slag, coarse cinders or other approved materials.

The effluent seeps out between the joints or through the holes in the disposal field piping. The best-designed disposal fields let air into the rock bed so that *aerobic* bacteria can decompose the effluent further. It's diffused and oxidized in the rock bed. The treated effluent adds nutrients to the soil under the disposal field.

Contamination from sewage is a real possibility in rural or urban areas where drinking water is taken from wells. Cross-connections can allow untreated sewage to enter the drinking water, spreading diseases such as cholera and typhoid.

Codes have strict rules on the installation of septic tanks, seepage pits, cesspools and disposal fields. They set standards for the size, tank construction, locations, types of materials and much more.

Septic tank and disposal field installation is an important part of a plumber's trade. Even if you don't intend to do much septic tank work, be familiar with the requirements. In any case, your plumbing examination will almost certainly include several questions similar to the questions in this chapter.

Q 10-1 The liquid capacity of all septic tanks for dwelling units, the Code specifies, is determined by the number:

(A) of bedrooms

(B) of persons

(C) of bathrooms

(D) and type of plumbing fixtures

Standard Plumbing Code

Uniform Plumbing Code

Answer: A

Code response: The liquid capacity of all septic tanks for dwelling occupancies is determined by the number of bedrooms or apartment units.

Answer: A

Code response: The liquid capacity of all septic tanks for dwelling occupancies is determined by the number of bedrooms or apartment units.

Q 10-2 According to Code, in designing the liquid capacity of septic tanks for a strip store shopping center, the determining factor in sizing is the:

(A) number of persons

(B) plumbing fixture units

(C) number of toilet rooms

(D) engineering data available

Standard Plumbing Code

Uniform Plumbing Code

Answer: B

Code response: In other building occupancies (commercial) the liquid capacity of septic tanks is determined by the number of plumbing fixture units.

Answer: B

Code response: In other building occupancies (commercial) the liquid capacity of septic tanks is determined by the number of plumbing fixture units.

Q 10-3 Where leaching beds are permitted in lieu of trenches, the Code specifies that the area of each such bed must be:

(A) 25 percent greater

(B) 25 percent less

(C) 50 percent greater

(D) 50 percent less

Standard Plumbing Code

Uniform Plumbing Code

Answer: C

Code response: Where leaching beds are permitted in lieu of trenches, the area of each such bed shall be at least 50 percent greater.

Answer: C

Code response: Where leaching beds are permitted in lieu of trenches, the area of each such bed shall be at least 50 percent greater.

Q 10-4 When the Code permits the installation of a disposal field, the minimum size of trench bottom has to be at least ____ square feet.

(A) 100

(B) 125

(C) 150

(D) 175

Standard Plumbing Code

Uniform Plumbing Code

Answer: C

Code response: When disposal fields are installed, a minimum of 150 square feet of trench bottom shall be provided.

Answer: C

Code response: When disposal fields are installed, a minimum of 150 square feet of trench bottom shall be provided.

Q 10-5 According to Code, the capacity of a septic tank and its drainage system:

(A) must be designed for maximum use

(B) must be designed for minimum use

(C) should have a back-up seepage pit

(D) must be limited by the soil structure classification

Standard Plumbing Code

Uniform Plumbing Code

Answer: D

Code response: The capacity of any one septic tank and its drainage system shall be limited by the soil structure classification.

Answer: D

Code response: The capacity of any one septic tank and its drainage system shall be limited by the soil structure classification.

Q 10-6 When the soil structure classification for a disposal field consists of coarse sand or gravel, the Code requires ____ square feet of leaching area for each 100 gallon capacity septic tank.

(A) 10

(B) 15

(C) 20

(D) 25

Standard Plumbing Code

Uniform Plumbing Code

Answer: C

Code response: 20 square feet of leaching area/100 gallons is required when soil structure classification consists of coarse sand or gravel.

Answer: C

Code response: 20 square feet of leaching area/100 gallons is required when soil structure classification consists of coarse sand or gravel.

Q 10-7 The maximum size permitted by Code for an apartment building septic tank is ____ gallons.

(A) 3,000

(B) 3,500

(C) 5,000

(D) 7,500

Standard Plumbing Code

Uniform Plumbing Code

Answer: D

Code response: The maximum septic tank size allowable is 7,500 gallons.

Answer: D

Code response: The maximum septic tank size allowable is 7,500 gallons.

Q 10-8 In order to determine the absorption qualities of questionable soils for a proposed disposal field, Code requires that the proposed site be:

(A) inspected by an engineer

(B) subjected to percolation tests

(C) first approved by local plumbing official

(D) inspected for porosity before back-filling

Standard Plumbing Code

Uniform Plumbing Code

Answer: B

Code response: In order to determine the absorption qualities of questionable soils, the proposed site shall be subjected to percolation tests.

Answer: B

Code response: In order to determine the absorption qualities of questionable soils, the proposed site shall be subjected to percolation tests.

Q 10-9 A 5,000 gallon septic tank has been approved for installation to serve a commercial building. To meet Code, the minimum leaching area has to be at least _____ square feet.

(A) 1,000
(B) 1,500

(C) 2,000
(D) 2,500

Standard Plumbing Code | Uniform Plumbing Code

Answer: C
Code response: Required square feet of leaching area/100 gallons septic tank capacity (5,000) is 40 square feet.

Answer: C
Code response: Required square feet of leaching area/100 gallons septic tank capacity (5,000) is 40 square feet.

Solution: 5,000 gal. (septic tank capacity) ÷ 100 gal. x 40 sq. ft. = 2,000 sq. ft. leaching area required

Q 10-10 Cesspools are not considered by most Codes to be an acceptable private sewage disposal system. When the Administrative Authority <u>does</u> permit a cesspool to be used, it:

(A) must be considered only as a temporary expedient
(B) may be used only in conjunction with an existing cesspool

(C) is limited to single family dwellings
(D) can be limited to receiving liquid waste only

Standard Plumbing Code | Uniform Plumbing Code

Answer: A
Code response: A cesspool shall be considered only as a temporary expedient, pending the construction of a public sewer.

Answer: A
Code response: A cesspool shall be considered only as a temporary expedient, pending the construction of a public sewer.

Q 10-11 When liquid wastes contain excessive amounts of grease, which could affect the proper operation of a private sewage disposal system, Code mandates:

(A) the installation of extra cleanouts
(B) an extra large disposal field

(C) the installation of an interceptor
(D) the installation of a backwater valve between the inlet and outlet tees

Standard Plumbing Code | Uniform Plumbing Code

Answer: C
Code response: Where appreciable amounts of indigestible waste are produced, a grease interceptor shall be installed as required by the Standard Plumbing Code.

Answer: C
Code response: When liquid wastes contain excessive amounts of grease, which may affect the operation of a private sewage disposal system, an interceptor for such wastes shall be installed.

Q. 10-12 The Code requires that all private sewage disposal systems be designed so that additional subsurface drainfields, equivalent to at least _____ percent of the required original system, may be installed if the original system cannot absorb all the sewage.

(A) 25
(B) 50

(C) 75
(D) 100

Standard Plumbing Code	Uniform Plumbing Code
Answer: D	**Answer: D**
Code response: All private sewage disposal systems shall be so designed that additional seepage pits or subsurface drainfields, equivalent to at least 100 percent of the required original system, may be installed if the original system cannot absorb all the sewage.	Code response: All private sewage disposal systems shall be so designed that additional seepage pits or subsurface drainfields, equivalent to at least 100 percent of the required original system, may be installed if the original system cannot absorb all the sewage.

Q. 10-13 Given: A 750 gallon septic tank having a minimum size disposal field of 150 square feet of trench bottom is to be installed on a particular lot. The source of water for the building is a domestic well. According to Code, the minimum horizontal distance between the septic tank and the well cannot be less than _____ feet.

(A) 25
(B) 50

(C) 75
(D) 100

Standard Plumbing Code	Uniform Plumbing Code
Answer: B	**Answer: B**
Code response: The minimum horizontal distance, in the clear, required from a water supply well and a septic tank is 50 feet.	Code response: The minimum horizontal distance, in the clear, required from a water supply well and a septic tank is 50 feet.

Q. 10-14 Given: A 1,000 gallon septic tank having a disposal field of 250 square feet of trench bottom is to be installed on a particular lot. The source of water for the building is a domestic well. The soil criterion is fine sand. According to Code, the minimum horizontal distance between the disposal field and the well must not be less than _____ feet.

(A) 25
(B) 50

(C) 75
(D) 100

Standard Plumbing Code	Uniform Plumbing Code
Answer: D	**Answer: D**
Code response: The minimum horizontal distance in the clear required from a water supply well and a disposal field is 100 feet.	Code response: The minimum horizontal distance in the clear required from a water supply well and a disposal field is 100 feet.

10-15 Given: A cesspool has been approved by the Administrative Authority as a sewage disposal facility for a single family residence. The source of water is a domestic well. The soil criteria are clay with considerable amounts of sand and gravel. The minimum horizontal distance between the cesspool and the well, according to Code, must not be less than _____ feet.

(A) 100

(B) 125

(C) 150

(D) 175

Standard Plumbing Code	Uniform Plumbing Code

Answer: C

Code response: The minimum horizontal distance in the clear required between a water supply well and a cesspool is 150 feet.

Answer: C

Code response: The minimum horizontal distance in the clear required between a water supply well and a cesspool is 150 feet.

10-16 Given: A rural single family residence is to be constructed on an extra large lot. A small stream crosses the property. The sewage disposal facilities consist of a septic tank and disposal field. The minimum horizontal distance between the septic tank and the stream, according to Code, can be no less than _____ feet.

(A) 50

(B) 75

(C) 100

(D) 150

Standard Plumbing Code	Uniform Plumbing Code

Answer: A

Code response: The minimum horizontal distance in the clear required from a stream and a septic tank is 50 feet.

Answer: A

Code response: The minimum horizontal distance in the clear required from a stream and a septic tank is 50 feet.

10-17 Given: A three bedroom, two bath house is to be constructed on an irregularly shaped lot. The septic tank's required measurements are 8'6" by 4'2". The architect shows the location of the tank on the east side of the building. According to Code, the minimum land space required for installation of the septic tank between the structure and the property line is:

(A) 8'0"

(B) 10'2"

(C) 12'4"

(D) 14'2"

Standard Plumbing Code	Uniform Plumbing Code

Answer: D

Code response: The minimum horizontal distance in the clear required from a structure is 5 feet and 5 feet from the property line.

Answer: D

Code response: The minimum horizontal distance in the clear required from a structure is 5 feet and 5 feet from the property line.

Solution: 5' from structure + 5' from property line + 4'2" tank width = 14'2" minimum land space required

Q 10-18 According to Code, a septic tank cannot be installed closer than _____ feet to the building water supply service.

(A) 5
(B) 8

(C) 10
(D) 12

Standard Plumbing Code	Uniform Plumbing Code
Answer: A Code response: Minimum horizontal distance in the clear required from on site domestic water service line and a septic tank is 5 feet.	**Answer: A** Code response: Minimum horizontal distance in the clear required from on site domestic water service line and a septic tank is 5 feet.

Q 10-19 The Code requires that drainage piping clear domestic water supply wells by at least 50 feet. This distance may be reduced to not less than 25 feet, provided <u>all but one</u> of the following drainage piping materials is used:

(A) cast iron standard weight
(B) copper type-L

(C) lead, extra heavy
(D) galvanized wrought iron

Standard Plumbing Code	Uniform Plumbing Code
Answer: D Code response: The distance may be reduced to not less than 25 feet when approved type metallic piping is installed.	**Answer: D** Code response: This distance may be reduced to not less than 25 feet when the drainage piping is constructed of materials approved for use within a building.

Note: The only drainage piping material listed above **not** approved for underground drainage piping within or outside a building is galvanized wrought iron. The correct answer is "D".

Q 10-20 To meet the Code, the maximum length of a disposal field drain line can be no more than _____ feet.

(A) 50
(B) 75

(C) 100
(D) 125

Standard Plumbing Code	Uniform Plumbing Code
Answer: C Code response: The maximum length of each distribution line shall be 100 feet.	**Answer: C** Code response: The maximum length of each distribution line shall be 100 feet.

Q 10-21 The minimum earth cover over drain lines in a disposal field, according to Code, must not be less than _____ inches.

(A) 12
(B) 14

(C) 16
(D) 18

Standard Plumbing Code	Uniform Plumbing Code
Answer: A Code response: The minimum depth of earth cover of lines in a disposal field is 12 inches.	**Answer: A** Code response: The minimum depth of earth cover of lines in a disposal field is 12 inches.

Q 10-22 Given: A two bedroom, one bath residence is to be constructed on sloping ground. The sewage disposal facilities consist of a 750 gallon septic tank and 150 square feet of disposal field. Referring to Figure 10-1, distance "X" shall be no less than _____ feet.

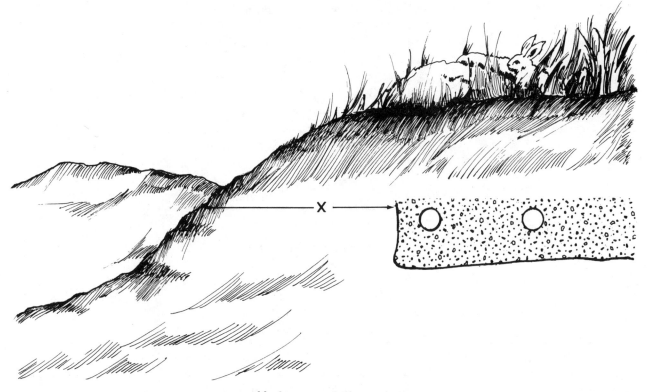

Underground disposal system
Figure 10-1

(A) 5
(B) 10

(C) 15
(D) 20

Standard Plumbing Code

Uniform Plumbing Code

Answer: C
Code response: When disposal fields are installed in sloping ground, the minimum horizontal distance between any part of the leaching system and ground surface shall be 15 feet.

Answer: C
Code response: When disposal fields are installed in sloping ground, the minimum horizontal distance between any part of the leaching system and ground surface shall be 15 feet.

Q 10-23 Using Figure 10-2, the minimum depth of filter material required by Code for distance "X" is most nearly _____ inches.

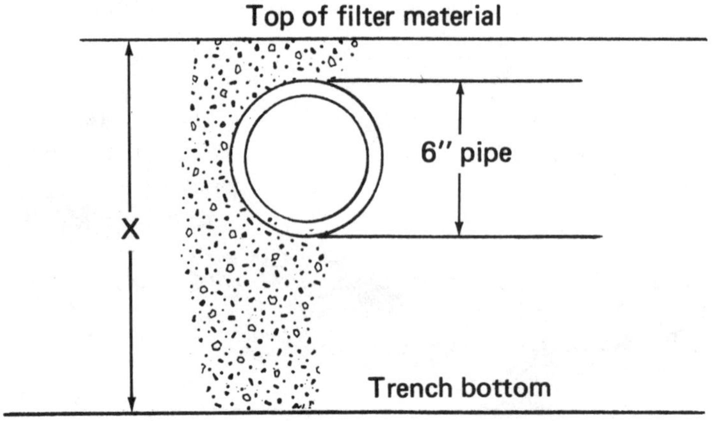

Top of filter material

6" pipe

X

Trench bottom

Filter material for drain line
Figure 10-2

(A) 8
(B) 12

(C) 16
(D) 20

Standard Plumbing Code

Uniform Plumbing Code

Answer: D
Code response: The minimum filter material under drain line shall be 12 inches and the minimum filter material over drain line shall be 2 inches.

Answer: D
Code response: The minimum filter material under drain line shall be 12 inches and the minimum filter material over drain line shall be 2 inches.

Solution: $\underset{\text{(under pipe)}}{12''}$ + 6" pipe + $\underset{\text{(over pipe)}}{2''}$ = 20" filter material

(Required for drain line as illustrated in Figure 10-2)

Q 10-24 When it is necessary to install a disposal field on sloping ground, to prevent excessive line slope, the Code requires that leach lines be:

(A) stepped
(B) laid level

(C) laid horizontal to vertical
(D) laid vertical to horizontal

Standard Plumbing Code

Uniform Plumbing Code

Answer: A
Code response: When necessary on sloping ground to prevent excessive line slope, leach lines shall be stepped.

Answer: A
Code response: When necessary on sloping ground to prevent excessive line slope, leach lines shall be stepped.

10-25 **When distribution lines of clay tile are used in a disposal field, Code mandates that they be laid:**

(A) in open trenches only
(B) in a disposal field only

(C) with open joints
(D) to maintain alignment

Standard Plumbing Code	Uniform Plumbing Code

Answer: C
Code response: Disposal field distribution lines when constructed of clay tile shall be laid with open joints.

Answer: C
Code response: Disposal field distribution lines when constructed of clay tile shall be laid with open joints.

10-26 **Three distribution lines constructed of perforated plastic pipe are to be used for a particular disposal field. Refer to the leaching bed illustrated in Figure 10-3. According to Code, the minimum width of disposal field excavation "X" is most nearly _____ feet.**

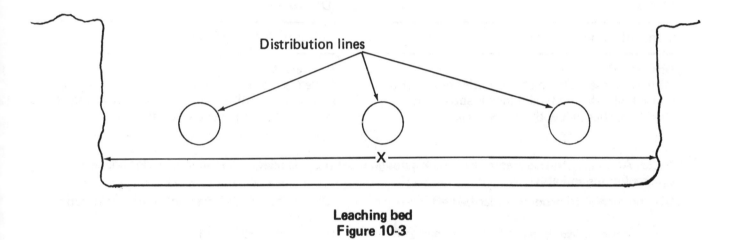

Distribution lines

Leaching bed
Figure 10-3

(A) 12
(B) 18

(C) 22
(D) 24

Standard Plumbing Code	Uniform Plumbing Code

Answer: B
Code response: Distribution drain lines in leaching beds shall not be more than 6 feet apart on centers and no part of the perimeter of the leaching bed shall be more than 3 feet from a distribution drain line.

Answer: B
Code response: Distribution drain lines in leaching beds shall not be more than 6 feet apart on centers and no part of the perimeter of the leaching bed shall be more than 3 feet from a distribution drain line.

Q 10-27 Where 2 or more drain lines are required to be installed in a leaching bed, Code requires the installation of a:

(A) pipe with watertight joints between the tank outlet and the drain lines

(B) approved backwater valve between the tank outlet and the drain lines

(C) pipe with sufficient openings for distribution of the effluent

(D) distribution box at the head of each disposal field

Standard Plumbing Code

Uniform Plumbing Code

Answer: D
Code response: An approved distribution box shall be constructed at the head of each disposal field.

Answer: D
Code response: An approved distribution box shall be installed at the head of each disposal field.

Q 10-28 According to Code, the inlet and outlet pipe (tee) for a septic tank must extend 4 inches above and at least _____ inches below the water surface.

(A) 4

(B) 8

(C) 12

(D) 18

Standard Plumbing Code

Uniform Plumbing Code

Answer: C
Code response: The inlet and outlet pipe (for a septic tank) shall extend 4 inches above and at least 12 inches below the water surface.

Answer: C
Code response: The inlet and outlet pipe (for a septic tank) shall extend 4 inches above and at least 12 inches below the water surface.

Q 10-29 According to Code, any septic tank's design must provide adequate space for sludge and scum accumulations and also:

(A) be such as to produce a clarified effluent

(B) support an earth load of not less than 200 pounds per square foot

(C) have one compartment to collect durable materials

(D) be constructed of concrete

Standard Plumbing Code

Uniform Plumbing Code

Answer: A
Code response: Septic tank's design shall be such as to produce a clarified effluent and shall provide adequate space for sludge and scum accumulations.

Answer: A
Code response: Septic tank's design shall be such as to produce a clarified effluent and shall provide adequate space for sludge and scum accumulations.

10-30 Septic tanks, according to Code, shall have a minimum of 2 compartments. The inlet compartment of a septic tank shall provide _____ of the total capacity of the tank.

(A) 1/4

(B) 1/3

(C) 1/2

(D) 2/3

Standard Plumbing Code

Answer: D

Code response: Septic tanks shall have a minimum of 2 compartments. The inlet compartment of any septic tank shall be not less than 2/3 of the total capacity of the tank.

Uniform Plumbing Code

Answer: D

Code response: Septic tanks shall have a minimum of 2 compartments. The inlet compartment of any septic tank shall be not less than 2/3 of the total capacity of the tank.

10-31 Where steel septic tanks are used, the Code requires a minimum wall thickness of no less than _____ gauge.

(A) 8

(B) 10

(C) 12

(D) 14

Standard Plumbing Code

Answer: C

Code response: The minimum wall thickness of any steel septic tank shall be No. 12 manufacturer's gauge.

Uniform Plumbing Code

Answer: C

Code response: The minimum wall thickness of any steel septic tank shall be No. 12 U.S. gauge.

10-32 Septic tanks are constructed with two compartments. According to Code, the inlet compartment for a 750 gallon septic tank must have a minimum liquid capacity of _____ gallons.

(A) 250

(B) 325

(C) 448

(D) 500

Standard Plumbing Code

Answer: D

Code response: Septic tanks shall have a minimum of 2 compartments. The inlet compartment of any septic tank shall be at least 500 gallons liquid capacity.

Uniform Plumbing Code

Answer: D

Code response: Septic tanks shall have a minimum of 2 compartments. The inlet compartment of any septic tank shall not be less than 500 gallons liquid capacity.

10-33 According to Code, a 4 bedroom single family residence would require the installation of a 1,200 gallon capacity septic tank. When designing a septic tank for a 7 bedroom single family residence, the liquid capacity would have to be increased to _____ gallons.

(A) 1,300
(B) 1,450

(C) 1,500
(D) 1,650

Standard Plumbing Code	Uniform Plumbing Code
Answer: D	**Answer: D**
Code response: For septic tank capacities exceeding 4 bedrooms for single family dwellings as listed in the table, "Capacity of Septic Tanks", add 150 gallons for each extra bedroom.	Code response: For septic tank capacities exceeding 4 bedrooms for single family dwellings as listed in the table, "Capacity of Septic Tanks", add 150 gallons for each extra bedroom.

Note: A 4 bedroom single family residence would require a 1,200 gallon septic tank. Add 150 gallons for each extra bedroom. (Three bedrooms would add 450 gallons = 1,650 gallons, total). Answer "D" is correct.

10-34 The architect calls upon your expertise as a plumbing contractor to help design a septic tank for a small apartment building. A careful review of the plans reveals that a total of 120 fixture units will be needed. After some calculating, you tell the architect that the minimum septic tank capacity acceptable by Code is _____ gallons.

(A) 3,250
(B) 3,500

(C) 3,550
(D) 4,000

Standard Plumbing Code	Uniform Plumbing Code
Answer: D	**Answer: D**
Code response: The minimum septic tank capacity in gallons up to 100 fixture units is 3,500 gallons. For each fixture unit over 100, 25 gallons per fixture unit must be added.	Code response: The minimum septic tank capacity in gallons up to 100 fixture units is 3,500 gallons. For each fixture unit over 100, 25 gallons per fixture unit must be added.

Solution:
$$\frac{\text{3,500 gal. tank}}{\text{(serving 100 FUs)}} + \frac{\text{25 gal. x 20 FUs}}{\text{(each FU requires 25 gal.)}} = \text{4,000 gal.}$$

3,500 gal. tank   +   500 gal.   = 4,000 gal. (septic tank capacity)

10-35 An 8-unit apartment building's minimum septic tank capacity is 3,000 gallons. In allowing for sludge storage capacity and the connection to the drainage system of 8 domestic food waste units, the minimum septic tank capacity required by Code is _____ gallons.

(A) 3,000
(B) 3,250

(C) 3,500
(D) 3,650

Standard Plumbing Code	Uniform Plumbing Code
Answer: A	**Answer: A**
Code response: Septic tank sizes in this table include sludge storage capacity and the connection of domestic food waste units without further volume increase.	Code response: Septic tank sizes in this table include sludge storage capacity and the connection of domestic food waste units without further volume increase.

10-36 According to Code, cleaning access to each septic tank must be provided. A septic tank having a first compartment of 12 feet in length would require one of the following:

(A) a manhole located over the inlet tee

(B) a manhole located over the inlet and one over the outlet tee

(C) a manhole located over the outlet tee

(D) a manhole located over the inlet and outlet tee and one located over the baffle wall

---

Standard Plumbing Code

Answer: B
Code response: One access manhole shall be located over the inlet and 1 access manhole shall be located over the outlet. Wherever a first compartment **exceeds 12 feet in length,** an additional manhole shall be provided over the baffle wall.

Uniform Plumbing Code

Answer: B
Code response: One access manhole shall be located over the inlet and 1 access manhole shall be located over the outlet. Wherever a first compartment **exceeds 12 feet in length,** an additional manhole shall be provided over the baffle wall.

10-37 Code requires that an air space be provided above the surface of the liquid level for the circulation of air within a septic tank. The minimum depth of this air space must be _____ inches.

(A) 7

(B) 8

(C) 9

(D) 10

Standard Plumbing Code

Answer: C
Code response: The free vent area total depth shall not be less than 9 inches greater than liquid depth.

Uniform Plumbing Code

Answer: C
Code response: The side walls (of a septic tank) shall extend at least 9 inches above the liquid depth.

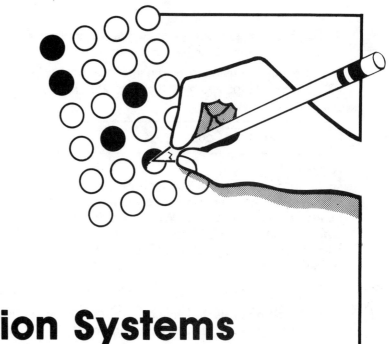

Chapter 11

Water Distribution Systems

Nearly every building intended for human habitation needs a water supply — wholesome potable water with adequate pressure and volume. Providing this is an important part of a plumber's job. Of course, your local health department will have to approve the source for the water supply. But water distribution is your job, once water leaves the public main.

The water distribution system must be designed to supply fixtures and equipment with enough water to ensure adequate performance and cleaning. Fixtures that receive too little volume or pressure don't work right and can be noisy in normal use.

The system you design must provide pressure of 8 pounds per square inch at all outlets and at all times, with the following exceptions:

1) At direct supply flush valves, the minimum required pressure is 15 psi.

2) At other equipment that needs more pressure, the minimum pressure is that needed to assure satisfactory performance.

The code regulates the type of piping materials used, installation methods, fittings, valves, protection of potable water supply and much more. The system must be maintained to prevent cross-connection, leakage and excessive waste of water.

Your plumbing exam will include many questions (and perhaps some piping diagrams) on the water distribution system. Review this section carefully to be well prepared at examination time.

Water Distribution Systems

Q 11-1 Of the following valves, the one Code-approved for a building water service line is a:

(A) globe valve
(B) gate valve

(C) angle valve
(D) needle valve

Standard Plumbing Code

Answer: B
Code response: A building control valve through which the water flows shall be equal to the cross-sectional area of the nominal size of the pipe in which it is installed.

Uniform Plumbing Code

Answer: B
Code response: A fullway valve controlling all outlets shall be installed on each building water service line.

Note: A gate valve is considered by Code to be a fullway valve and is required on a water service pipe to the building.

Q 11-2 According to Code and established plumbing standards, the one of the following fittings and valves that offers the least resistance to the flow of water under the same or similar conditions is a:

(A) gate valve
(B) swing check valve

(C) 90-degree copper ell
(D) 90-degree galvanized ell

Standard Plumbing Code

Answer: A
Code response: A gate valve shall be equal to the cross-sectional area of the nominal size of the pipe in which it is installed.

Uniform Plumbing Code

Answer: A
Code response: A fullway valve controlling all outlets shall be installed on each water service line.

Note: A gate valve (fullway valve) is considered by Code to be less resistant to the flow of water than any other fitting or valve listed above.

Q 11-3 The minimum diameter acceptable by Code for a building water service pipe manufactured from material other than lead or brass is _____ inch.

(A) 1/2
(B) 3/4

(C) 7/8
(D) 1

Standard Plumbing Code

Answer: B
Code response: Water service supply piping shall not be less than ¾-inch in diameter.

Uniform Plumbing Code

Answer: B
Code response: No building water supply pipe shall be less than ¾-inch in diameter.

Q 11-4 According to Code, of the following hot water temperature ranges, the one required for a commercial dishwashing machine is:

(A) 110° - 120° F.
(B) 130° - 140° F.

(C) 170° - 180° F.
(D) 190° - 200° F.

Standard Plumbing Code	Uniform Plumbing Code
Answer: C Code response: Dishwashing machines not installed in private residential units shall be provided with water at 180° F.	**Answer:** ____ Code response: None

Note: Many local Codes do not address temperature requirements for commercial dishwashing machines. Jurisdiction is held by the health department, and they usually require a minimum temperature of 180° F. for water to a commercial dishwashing machine. Answer "C" is the acceptable answer.

Q 11-5 An irrigation system is connected to the building water service piping. Code requires that an atmospheric vacuum breaker be installed on the irrigation branch pipe connection. The purpose of the vacuum breaker is to:

(A) limit the flow of water to the sprinkler heads

(B) equalize the water pressure for the two systems

(C) provide a safety precaution to prevent cross-connection

(D) prevent excessive pressure in the irrigation system

Standard Plumbing Code	Uniform Plumbing Code
Answer: C Code response: Lawn sprinkling systems shall be equipped with an approved vacuum breaker to protect against contamination of the potable water system.	**Answer: C** Code response: Lawn sprinkling systems shall be equipped with an approved vacuum breaker to protect against contamination of the potable water system.

Q 11-6 Given: The available pressure at the water meter is 40 psi, and the maximum length of run is 82 feet. The Code-accepted minimum diameter of the water supply branch to a stall urinal is ____ inch.

(A) 1/2
(B) 5/8

(C) ¾
(D) 1

Standard Plumbing Code	Uniform Plumbing Code
Answer: C Code response: The supply control for a stall urinal shall be ¾-inch in diameter.	**Answer: C** Code response: A stall urinal equals 5 fixture units and the supply outlet shall be ¾-inch.

11-7 Given: A 3-inch exposed copper water main for a 3-story building is hung to the ceiling of the first floor parking garage. In accordance with good plumbing practice, adequate hangers must be provided. The Code-mandated minimum spacing between hangers is _____ feet.

(A)   8   (C)   10
(B)   9   (D)   12

Standard Plumbing Code	Uniform Plumbing Code
Answer: C Code response: Horizontal copper tubing shall be supported at 10-foot intervals for piping 2 inches and larger.	**Answer: C** Code response: Horizontal copper tubing shall be supported at 10-foot intervals for piping 2 inches and larger in diameter.

In answering Questions 11-8 through 11-16, refer to the valve and fitting symbols shown in Figure 11-1. Read each question carefully before selecting your answer.

11-8 You are installing a 1-inch galvanized steel water service line to the building. The Code requires that a house valve be installed to control the building's water supply. The valve you would select from Figure 11-1 for this particular installation is number _____ .

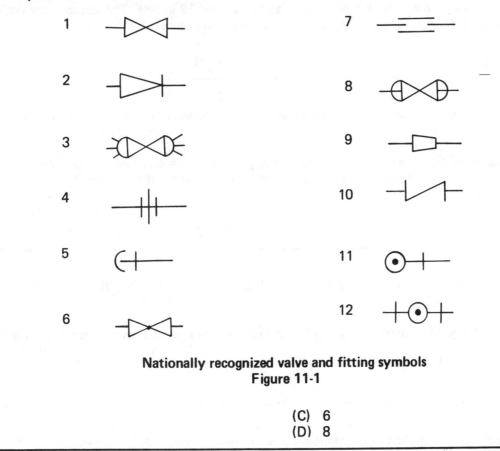

Nationally recognized valve and fitting symbols
Figure 11-1

(A)   1   (C)   6
(B)   3   (D)   8

Answer: A

Note: Symbol Number 1 illustrates a threaded or screwed gate valve. Answer "A" is correct.

Q 11-9 The Code requires that a gate valve be installed near a hot water heater in the cold water supply pipe. The supply pipe is ¾-inch type-L copper. The valve you would select from Figure 11-1 is numbered ____ .

(A) 1   (C) 6

(B) 3   (D) 8

Answer: D

Note: Symbol Number 8 illustrates a soldered gate valve. Answer "D" is correct.

Q 11-10 The Code requires the installation of watertight valves to control the hot and cold water to a restaurant pot sink faucet. The fixture supply piping is ½-inch galvanized steel pipe. The type of valve you would select for this job, from Figure 11-1, is numbered ____.

(A) 2   (C) 6

(B) 3   (D) 8

Answer: C

Note: Symbol Number 6 illustrates a threaded or screwed globe valve. Globe valves are considered to be watertight by Code. Answer "C" is correct.

Q 11-11 There is a need to extend a 1-inch galvanized steel pipe 10 inches. The appropriate fitting necessary to accomplish this job is numbered ____ as illustrated in Figure 11-1.

(A) 4   (C) 7

(B) 5   (D) 9

Answer: C

Note: Symbol Number 7 illustrates a threaded or screwed coupling. Answer "C" is correct.

Q 11-12 A black steel pipe fire sprinkler system is being installed in a commercial building. The engineer specifies that all joints are to be welded. The correct valve to use for this system is numbered ____, as illustrated in Figure 11-1.

(A) 1   (C) 6

(B) 3   (D) 8

Answer: B

Note: Symbol Number 3 illustrates a gate valve with welded ends. Answer "B" is correct.

Q 11-13 Of the fittings illustrated in Figure 11-1, the one representing an elbow turned down is numbered ____.

(A) 5   (C) 9

(B) 7   (D) 11

Answer: A

Note: Symbol Number 5 illustrates an elbow turned down. Answer "A" is correct.

11-14 Of the fittings illustrated in Figure 11-1, the one representing an elbow turned up is numbered ____.

(A) 5   (C) 2
(B) 11   (D) 12

Answer: B

Note: Symbol Number 11 illustrates an elbow turned up. Answer "B" is correct.

11-15 You are installing gas piping in a take-out restaurant using 1-inch diameter galvanized steel pipe. You need to reduce a run of pipe to ¾-inch. The correct fitting, according to Code, is numbered ____ as illustrated in Figure 11-1.

(A) 2   (C) 9
(B) 4   (D) 12

Answer: A

Note: Symbol Number 2 illustrates a reducer fitting acceptable by Code in a gas piping system. Answer "A" is correct.

11-16 To reduce the branch outlet of a ¾-inch galvanized tee to ½-inch in a water supply system, the Code permits the use of the fitting numbered ____, as illustrated in Figure 11-1.

(A) 4   (C) 7
(B) 5   (D) 9

Answer: D

Note: Symbol Number 9 illustrates a screwed bushing. Codes permit the use of bushings to reduce fitting sizes in water piping systems. Answer "D" is correct.

11-17 The type of threads required by Code on galvanized steel used in a water pipe system is:

(A) Briggs standard pipe threads   (C) WWP-401 regular pipe threads
(B) standard taper pipe threads   (D) WWT-791 standard pipe threads

Standard Plumbing Code   Uniform Plumbing Code

Answer: B   **Answer: B**
Code response: Threaded joints shall conform   Code response: Threads on iron pipe size (I.P.S.)
to American National Taper pipe threads   pipe shall be standard taper pipe threads.

Note: Although both reference Codes provide differing expressions for types of threads, they actually mean the same. Threads on all threadable pipe used in plumbing systems are standard taper pipe thread.

11-18 Any large plumbing job would reflect all four piping symbols illustrated below. In scaling and sizing the hot water return line, the piping symbol you would select is:

(A) —————— — ———— — ———— — — ———

(B) —————— — — ———— — — ———— — — ——

(C) —————— — — ———— — — ———— — — ———

(D) —

Answer: C

Note: In the piping symbols shown above,

"A" is cold water,
"B" is hot water,
"C" is hot water return,
"D" is vent piping.

Answer "C" is correct.

11-19 The plumber, according to Code, is responsible for providing adequate protection to prevent the water system from possible:

(A) temperature variation

(B) stresses

(C) back-siphonage

(D) pressure drops

Standard Plumbing Code

Uniform Plumbing Code

Answer: C
Code response: Adequate protection shall be provided to prevent possible backflow of an unsafe fluid into a safe water system.

Answer: C
Code response: No installation of potable water supply piping or part thereof shall be made in such a manner that it is possible for unclean fluid to enter any portion of such piping.

11-20 The one of the following uses **not** requiring potable water (as defined by Code) is water used for:

(A) land irrigation

(B) culinary purposes

(C) domestic households

(D) private swimming pools

Standard Plumbing Code

Uniform Plumbing Code

Answer: A
Code response: Potable water shall be used for drinking, culinary, and domestic purposes.

Answer: A
Code response: Potable water is water which is satisfactory for drinking, culinary and domestic purposes.

11-21 Water pressure in public water mains averages approximately 45 to 50 pounds per square inch. The Code considers that pressure in public water mains is often excessive and requires that a pressure-reducing valve be installed when the pressure exceeds ____ pounds per square inch.

(A) 65

(B) 70

(C) 75

(D) 80

Standard Plumbing Code

Uniform Plumbing Code

Answer: D
Code response: When water pressure exceeds 80 psi, a water pressure-reducing valve shall be installed.

Answer: D
Code response: Where local water pressure is in excess of 80 pounds per square inch (psi), an approved type pressure regulator shall be installed.

11-22 When the residual pressure in a water main is not adequate to provide the minimum water pressure for plumbing fixtures to function properly at the highest water outlet, the Code mandates that the plumber install:

(A) the next larger pipe size for the building water service line

(B) a water pressure pump

(C) an approved engineered water hammer arrester on top of most pipes

(D) a pressure tank in the building water service line

Standard Plumbing Code	Uniform Plumbing Code
Answer: B Code response: If the residual pressure in the system is below the minimum allowable at the highest water outlet, an automatically controlled pump shall be installed.	**Answer: B** Code response: Whenever the water pressure in the main will not provide a water pressure of at least 15 psi, a pump shall be installed.

11-23 According to Code, if the highest group of fixtures contains flushometer valves, the minimum pressure required for the fixtures to function properly would be ___ psi.

(A) 8
(B) 11
(C) 15
(D) 18

Standard Plumbing Code	Uniform Plumbing Code
Answer: C Code response: If the highest group of fixtures contains flush valves, the pressure for the group should not be less than 15 psi.	**Answer: C** Code response: If the highest group of fixtures contains flushometer valves, the pressure for the group should not be less than 15 psi.

11-24 To meet Code, a three-story apartment building having tank-type flush water closets must have a minimum pressure of ___ psi on the third floor.

(A) 8
(B) 11
(C) 15
(D) 18

Standard Plumbing Code	Uniform Plumbing Code
Answer: A Code response: If the highest group of fixtures contains flush-tank supplies, the available pressure may be not less than 8 psi.	**Answer: A** Code response: If the highest group of fixtures contains flush tank supplies, the available pressure may not be less than 8 psi.

Q 11-25 A lawn sprinkler system receives its source of water from the building supply line. The Code requires the installation of a vacuum breaker to prevent reverse flow into the potable water supply. According to Code, such vacuum breaker must be installed:

(A) in an accessible valve box

(B) at least 6 inches above the highest sprinkler head

(C) a minimum of 4 inches below the highest sprinkler head

(D) to isolate the sprinkler system from other water sources

Standard Plumbing Code

Answer: B
Code response: Atmospheric-type vacuum breakers at least 6 inches above the level of the highest sprinkler head shall be installed.

Uniform Plumbing Code

Answer: B
Code Response: The vacuum breaker shall be installed at least 6 inches above the surrounding ground.

Q 11-26 To meet Code requirements in Vero Beach, Florida, hose bibbs installed on a single family residence must:

(A) be a maximum of ½-inch pipe diameter in size

(B) exit from the building wall a minimum of 12 inches above grade.

(C) be protected from freezing

(D) be protected by an approved non-removable type backflow prevention device.

Standard Plumbing Code

Answer: D
Code response: Hose bibbs and lawn hydrants shall be protected by an approved back-siphonage backflow preventer.

Uniform Plumbing Code

Answer: D
Code response: Hose bibbs and lawn hydrants shall be protected by an approved non-removable type backflow prevention device.

Q 11-27 A 10-foot section of cold water piping is unavoidably installed in the exterior wall of an apartment building in Crossville, Tennessee. The plumbing inspector turns down the plumber's work because he:

(A) used a pipe size larger than required by Code

(B) had not graded the pipe to drain dry

(C) failed to allow for contraction and expansion of the pipe

(D) did not make adequate provisions to protect it from freezing

Standard Plumbing Code

Answer: D
Code response: Water pipe shall not be installed in outside walls where it is subjected to freezing temperature, unless adequate provision is made to protect it from freezing.

Uniform Plumbing Code

Answer: D
Code response: No water pipe shall be installed in an exterior wall, unless adequate provision is made to protect such pipe from freezing.

11-28 When connecting a building cold water supply to a steam boiler, the cold water connection must be protected. According to Code, the installation must include:

(A) in addition to an approved gate valve, a brass seat swing check valve

(B) an approved backflow prevention device

(C) an aspirator device

(D) a double check type valve with strainer

Standard Plumbing Code

Uniform Plumbing Code

Answer: B
Code response: Water supply to a steam boiler shall have installed a backflow preventer device in the water supply line to prevent back-siphonage and backflow from the heating system into the potable supply line.

Answer: B
Code response: Steam boiler connections shall be protected by an approved backflow prevention device.

11-29 The water pipe sizing procedure, according to Code, is based on:

(A) static pressure loss or gain

(B) pressure required at fixture to produce required flow

(C) the friction or pressure loss through water meter

(D) a system of pressure requirements and losses

Standard Plumbing Code

Uniform Plumbing Code

Answer: D
Code response: The water pipe sizing procedure is based on a system of pressure requirements and losses.

Answer: D
Code response: Size of potable water piping shall be determined by the pressure requirements and losses.

11-30 After cutting the desired length of type-L copper water tubing to fit between an elbow and a tee, the Code requires that:

(A) the outside be cleaned bright

(B) the joints and pipe be properly fluxed

(C) a 50-50 solder be used to make the joints

(D) the tubing be returned to full bore

Standard Plumbing Code

Uniform Plumbing Code

Answer: D
Code response: All burrs shall be removed and the tubing shall be returned to full bore.

Answer: D
Code response: Burred ends shall be reamed to the full bore of the tube.

11-31 The Code usually requires that appropriate fittings be used in direction changes for water piping systems. Changes in direction in <u>copper</u> tubing may be made with bends, <u>if</u> such bends are made with bending equipment which does not:

(A) cause tubing to become brittle
(B) weaken the walls of the tubing
(C) promote deformity of the tubing
(D) create difficulties in installation of tubing

Standard Plumbing Code

Answer: C
Code response: Changes in direction in copper tube may be made with bends, provided that such bends are made by use of forming equipment which does not deform or create a loss in cross-sectional area of the tube.

Uniform Plumbing Code

Answer: C
Code response: Changes in direction in copper tubing may be made with bends, provided that such bends are made with bending equipment which does not deform or create a loss in cross-sectional area of the tubing.

11-32 A 3-inch vertical water main is being installed in a six-story building. The piping material is type-L copper. There is a separation of 9 feet between floors. According to Code, a minimum of _____ supports would be needed to properly support the main.

(A) 3
(B) 6
(C) 9
(D) 12

Standard Plumbing Code

Answer: B
Code response: Copper tube shall be supported at each story for piping 1½-inches and over.

Uniform Plumbing Code

Answer: B
Code response: Copper tubing shall be supported at each story or at maximum intervals of 10 feet.

11-33 You are designing a water system for a four-plex building. The water pressure fluctuates between a low of 30 psi in late afternoon to a high of 45 psi in mid-morning. The Code mandates the water piping system be designed for water pressure of no less than _____ psi.

(A) 30
(B) 35
(C) 40
(D) 45

Standard Plumbing Code

Answer: A
Code response: When the street main has a wide fluctuation in pressure, the water distribution system shall be designed for minimum pressure available.

Uniform Plumbing Code

Answer: A
Code response: In localities where there is a fluctuation of pressures in the main throughout the day, the water piping systems shall be designed on the basis of the minimum pressure available.

Q 11-34 Flush tanks shall be equipped with approved ballcocks. The ballcock shall be equipped with a vacuum breaker. The Code mandates that such vacuum breaker be located:

(A) on tank water supply line to prevent backflow

(B) on the overflow tube, 1 inch above the tank water level

(C) as an integral part of the ballcock, 1 inch above the tank water level

(D) as an integral part of the ballcock, 1 inch above the overflow pipe

Standard Plumbing Code

Uniform Plumbing Code

Answer: D
Code response: The ballcock shall be installed with the critical level of the vacuum breaker at least 1 inch above the full opening of the overflow pipe.

Answer: D
Code response: The ballcock shall be installed with the critical level of the vacuum breaker at least 1 inch above the full opening of the overflow pipe.

Q 11-35 To ensure that vacuum breakers perform satisfactorily under their intended service condition, the Code requires that they:

(A) be maintained in good working condition

(B) are replaced with new devices once each year

(C) are inspected frequently for leakage

(D) be installed a maximum of 6 inches above the overflow rim of the fixture they serve.

Standard Plumbing Code

Uniform Plumbing Code

Answer: A
Code response: All devices for the prevention of backflow shall be maintained in good working condition.

Answer: A
Code response: All devices installed in a potable water supply system for protection against backflow shall be maintained in good working condition.

Q 11-36 Pipe is made in three weights: standard, extra heavy, and double extra heavy. When comparing 4-inch standard weight black steel pipe with 4-inch double extra heavy galvanized steel pipe, it can be generally said that:

(A) the weight per foot of one is twice that of the other

(B) the nominal size is the same for each pipe

(C) the outside diameters are the same

(D) the same threading dies cannot be used for both pipes

Standard Plumbing Code

Uniform Plumbing Code

Answer: C

Answer: C

Note: According to Code and the manufacturer, the outside diameters of threadable pipe are alike, so that the same threading dies will fit all three weights of pipe.

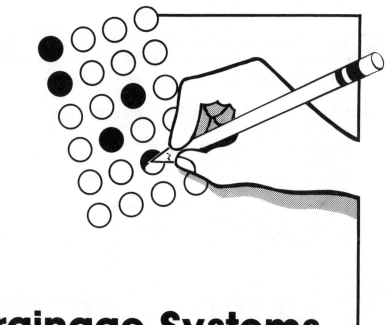

Chapter 12

Storm Water Drainage Systems

The storm drainage system collects rain water and melted snow and carries them to a point where they may be legally discharged. Storm water drainage includes roof drains, area drains, catch basins, gutters, leaders, building storm drains, building storm sewers and ground surface storm sewers.

Rain water not properly collected and disposed of can become both a nuisance and a health hazard. Collection and disposal of storm water is now considered one of the primary functions of the plumbing drainage system.

In the past, many cities situated near large lakes, rivers, or the ocean, used a combined sewer system to convey both storm water and sewage. This is no longer permitted by code. Exceptions can be made in hardship cases with prior approval from the local authority. With this approval, new construction can discharge into a dual storm water and sewage system. Otherwise, the code now requires separate sewage and storm water collection systems.

On large jobs, professional engineers usually size storm water drainage pipes and determine the method of disposal. Their calculations are based on two factors: the square feet of the water collecting area (roofs, parking lots, etc.) and the maximum anticipated rainfall rate in inches for any one hour.

As a plumber, you should be familiar with the principles involved in planning storm drainage systems. The journeyman and master plumber exam will almost certainly include several questions on storm drainage systems.

Q 12-1 The Plumbing Code requires that a strainer be provided for each building roof drain. Such strainers must extend not less than _____ inches above the surface of the roof immediately adjacent to the roof drain.

(A) 2
(B) 3

(C) 4
(D) 5

Standard Plumbing Code

Uniform Plumbing Code

Answer: C
Code response: Such strainers shall extend not less than 4 inches above the surface of the roof immediately adjacent to the roof drain.

Answer: C
Code response: Roof drains shall be equipped with strainers extending not less than 4 inches above the surface of the roof immediately adjacent to the drain.

Q 12-2 Given: The projected roof area of a building is 4,600 square feet. The maximum rainfall is 4 inches. The rain leader pipe, according to Code, should be no less than _____ inches.

(A) 2½
(B) 3

(C) 4
(D) 5

Standard Plumbing Code

Uniform Plumbing Code

Answer: C
Code response: See Table 1506.1 which shows that a 4-inch rain leader would be required for a projected roof area of 4,600 square feet and a maximum rainfall of 4 inches.

Answer: C
Code response: See Table D-1 which shows that a 4-inch rain leader would be required for a projected roof area of 4,600 square feet and a maximum rainfall of 4 inches.

Q 12-3 According to Code, roof drain strainers must have a minimum inlet area _____ times the pipe to which it is connected.

(A) 1
(B) 1½

(C) 2
(D) 2½

Standard Plumbing Code

Uniform Plumbing Code

Answer: B
Code response: Roof drain strainers shall have an available inlet area, above roof level, of not less than 1½ times the area of the leader to which the drain is connected.

Answer: B
Code response: Roof drain strainers shall have a minimum inlet area 1½ times the pipe to which it is connected.

Q 12-4 In sizing roof drains and storm drainage piping, _____ percent of a vertical wall which diverts rain water to the roof must be added to the projected roof area when calculating the Code-required storm drainage piping.

(A) 50   (C) 40
(B) 25   (D) 60

Standard Plumbing Code | Uniform Plumbing Code

Answer: A
Code response: In sizing roof drains and storm drainage piping, ½ of the area of any vertical wall which diverts rain water to the roof shall be added to the projected roof area for inclusion in calculating the required size of vertical leaders and horizontal storm drainage piping.

Answer: A
Code response: In sizing roof drains and storm drainage piping where a vertical wall projects above a roof so as to permit storm water to drain to the roof area below, add 50 percent of the wall area to the projected roof area to calculate required storm drainage piping sizes.

Q 12-5 Given: It is necessary to install a hand sink a considerable distance from the nearest building sanitary drainage system. A 4-inch PVC Schedule 40 rain leader is conveniently located. Since the waste from the hand sink will not contain fecal matter, the Code:

(A) will permit an indirect connection to the rain leader
(B) will permit a direct connection to the rain leader
(C) requires an indirect connection to the sanitary system
(D) requires a direct connection to the sanitary system

Standard Plumbing Code | Uniform Plumbing Code

Answer: D
Code response: All plumbing fixtures used to receive or discharge liquid wastes shall be connected properly to the drainage system. Rain water conductor pipes shall not be used as waste pipes.

Answer: D
Code response: All plumbing fixtures used to receive or discharge liquid wastes shall be connected properly to the drainage system. Rain water piping shall not be used as waste pipes.

Q 12-6 Given: You are installing a rain water leader in a parking garage serving a six-story office building. It is firmly secured to a column supporting the upper floors. According to Code:

(A) it should not be secured to the supporting column
(B) it should be relocated to the inside wall
(C) it must be protected
(D) the material for the rain water leader must be no less than cast iron pipe

Standard Plumbing Code | Uniform Plumbing Code

Answer: C
Code response: Rain water conductors installed in locations where they may be exposed to damage shall be protected.

Answer: C
Code response: Rain water piping installed in locations where it may be subject to damage shall be protected.

12-7 According to Code, which one of the following answers is <u>not</u> correct?

(A) Roof drains shall be constructed of cast iron

(B) Roof drains must be constructed of corrosion resisting material

(C) Roof drains shall be constructed of copper

(D) Roof drains may be constructed of corrosion type material, since they receive clear water wastes only

---

Standard Plumbing Code | Uniform Plumbing Code

Answer: D
Code response: Roof drains shall be of an approved corrosion-resistant material.

Answer: D
Code response: Roof drains shall be of cast iron, copper, lead or other corrosion resisting material.

12-8 Rain water piping installed within a vent or shaft of a building, according to Code, shall <u>not</u> be:

(A) Schedule 40 PVC plastic piping

(B) 26-gauge galvanized sheet metal piping

(C) cast iron piping

(D) galvanized steel piping

Standard Plumbing Code | Uniform Plumbing Code

Answer: B
Code response: Rain water conductors run in vent or pipe shafts shall be cast iron, galvanized steel, Schedule 40 plastic, or other approved materials.

Answer: B
Code response: Rain water piping placed within a vent or shaft shall be cast iron, Schedule 40 PVC, galvanized steel, or other approved materials.

Note: 26-gauge galvanized sheet metal piping is approved only for exterior use. Correct answer is "B".

12-9 You have installed the underground storm drainage system within a building using Schedule 40 PVC plastic pipe. According to Code, the underground portion must be tested with at least a _____ -foot head of water.

(A) 5

(B) 7

(C) 10

(D) 12

Standard Plumbing Code | Uniform Plumbing Code

Answer: C
Code response: Leaders within a building shall be tested by water as required by the methods of testing drainage and vent systems.

Answer: C
Code response: Rain water piping when concealed within the construction of the building shall be tested in conformity with the provisions of this Code for testing drain, waste and vent systems.

Note: The Code states that no section of the drainage and vent system shall be tested with less than a 10-foot head of water. Answer "C" is correct.

12-10 Given: A job calls for 2 roof deck strainers on a sun deck. The Code would require that such drains have a minimum inlet area _____ times the area of the pipe to which the drain is connected.

(A) 1
(B) 1½

(C) 2
(D) 2½

Standard Plumbing Code

Uniform Plumbing Code

Answer: C
Code response: Roof drain strainers for use on sun decks shall have an available inlet area not less than 2 times the area of the leader to which the drain is connected.

Answer: C
Code response: Roof deck strainers for use on sun decks shall have an inlet area not less than 2 times the area of the pipe to which the drain is connected.

12-11 Because of the limited fall available for the installation of a building storm drainage system, a fall of 1/8-inch per foot is used. If the maximum rainfall is 4 inches per hour and the roof area is 27,350 square feet, the Code-required minimum pipe size for the building storm sewer is _____ inches.

(A) 8
(B) 10

(C) 12
(D) 15

Standard Plumbing Code

Uniform Plumbing Code

Answer: C
Code response: See Table 1506.2 which shows that the maximum rainfall in inches per hour is 4 inches. The building roof is 27,350 square feet. Horizontal piping installed at 1/8-inch fall per foot shall be 12 inches in diameter.

Answer: C
Code response: See Table D-2 which shows that the maximum rainfall in inches per hour is 4 inches. The building roof is 27,350 square feet. Horizontal piping installed at 1/8-inch fall per foot shall be 12 inches in diameter.

12-12 Which of the following statements is most nearly correct, according to Code?

(A) Rain water piping passing under a wall needs no protection from breakage
(B) Storm water drainage piping passing through corrosive material must be protected against internal corrosion

(C) Rain water piping passing through a wall needs to be protected from breakage
(D) Provisions shall be made for the contraction of rain water leaders

---

Standard Plumbing Code

Uniform Plumbing Code

Answer: C
Code response: Pipes passing under or through walls shall be protected from breakage.

Answer: C
Code response: All piping passing under or through walls shall be protected from breakage.

12-13 Which of the following statements is **most nearly correct**, according to Code?

(A) The minimum grade for horizontal rain water piping acceptable by Code is 1/16-inch fall per foot

(B) The minimum size rain leader acceptable by Code is 2 inches in diameter

(C) All Codes base their storm drainage pipe sizes on 4 inches of rainfall per hour

(D) When sizing rain leaders, projected roof areas need not be considered

---

Standard Plumbing Code

Answer: B
Code response: Minimum size of leader is 2 inches.

Uniform Plumbing Code

Answer: B
Code response: Minimum size of leader is 2 inches.

Note: See Table 1506.1 **Standard Plumbing Code** and Table D-1 **Uniform Plumbing Code**

12-14 Which of the following statements is **most nearly correct**, according to Code?

(A) Vertical leaders may be decreased in size before passing through the roof

(B) Vertical leaders shall be sized for the maximum projected roof area

(C) Rain leaders shall be sized in accordance with the type roof drains required

(D) Under no circumstances shall rain leaders exceed the minimum pipe sizes as required by Code

---

Standard Plumbing Code

Answer: B
Code response: Vertical leaders shall be sized for the maximum projected roof area.

Uniform Plumbing Code

Answer: B
Code response: Vertical rain water piping shall be sized in accordance with Table D-1 and is based on a given roof area (measured in square feet).

12-15 Copper tubing is made in four weights, called types. According to Code, which of the following statements is **most nearly correct**?

(A) DWV copper pipe comes in approximately 20-foot lengths

(B) The minimum weight for rain leaders is no less than type-L copper

(C) Rain water leaders made of copper pipe need not be water tested

(D) Solder used to make joints in a copper rain water piping system shall be no less than 60-40

Standard Plumbing Code

Answer: A
Code response: None

Uniform Plumbing Code

Answer: A
Code response: None

Note: Hard copper tubing is manufactured in 20-foot lengths. Codes generally do not address or require copper tubing to be a specified length.

Q **12-16 Given:** An engineer designs a building storm drainage system in type-M copper tubing. The building height is 6 stories. The rainfall in the area is based on 4 inches per hour. He calculates that he will need two 4-inch roof drains and one 3-inch roof drain. According to Code, the roof area to be drained would be most nearly _____ square feet.

(A) 6,600
(B) 9,200

(C) 11,400
(D) 12,800

Standard Plumbing Code

Uniform Plumbing Code

Answer: C
Code response: Based on a rainfall of 4 inches per hour, vertical rain leaders shall be sized according to Table 1506.1.

Answer: C
Code response: Based on a rainfall of 4 inches per hour, leaders shall be sized according to Table D-1.

Note: Each 4-inch leader will drain 4,600 square feet and a 3-inch leader 2,200 square feet of projected roof area, according to Table 1506.1 and Table D-1. Answer "C" is correct.

Q **12-17 Refer again to Question 2-16.** According to Code, to properly support all 3 leaders, you will need to install the following number and sizes of pipe floor clamps:

(A) ten 3-inch and twenty 4-inch clamps
(B) five 3-inch and ten 4-inch clamps

(C) six 3-inch and ten 4-inch clamps
(D) six 3-inch and twelve 4-inch clamps

Standard Plumbing Code

Uniform Plumbing Code

Answer: D
Code response: Copper tube shall be supported at each story for piping 1½ inches and over.

Answer: D
Code response: Copper tubing shall be supported at each story.

Q **12-18** Code mandates that roof leaders penetrating the roof of a building meet <u>one</u> of the following requirements:

(A) The hole cut to receive leader pipe shall not be more than ¼-inch larger than the outside diameter of the leader pipe
(B) The pipe shall be increased one pipe size larger to prevent frost or snow closure

(C) Joints at the roof around leader pipes shall be made watertight

(D) Each leader shall terminate not less than 10 feet from any air intake or vent shaft

Standard Plumbing Code

Uniform Plumbing Code

Answer: C
Code response: The connection between roofs and roof drains which pass through the roof and into the interior of the building shall be made watertight.

Answer: C
Code response: Roof drains passing through the roof into the interior of a building shall be made watertight at the roof line.

12-19 Code requires that large open parking decks have drains to receive rain water and convey it to a suitable disposal area. Therefore, which of the following statements is most nearly correct?

(A) The parking deck drainage system shall connect into the nearest leader pipe from the principal building

(B) All parking deck drains are generally required to have only flat surface type strainers

(C) All deck type drains shall be no less than 4 inches in diameter

(D) Parking deck drains shall first drain through a special separator to remove harmful ingredients before connecting into the storm water drainage system

---

Standard Plumbing Code

Answer: B
Code response: Roof drain strainers for use on parking decks may be of the flat surface type.

Uniform Plumbing Code

Answer: B
Code response: Roof deck strainers for use on parking decks may be of an approved flat-surface type.

12-20 Given: An engineer wishes to use a 3-inch leader to collect rain water from a portion of the roof. The rainfall is 4 inches per hour. In his calculations, the maximum square footage of roof area which he can use for one leader is ____ square feet.

(A) 1,700
(B) 2,200

(C) 3,100
(D) 4,400

Standard Plumbing Code

Answer: B
Code response: See Table 1506.1 which shows that based on a 4-inch per hour rainfall, a 3-inch leader will serve a maximum projected roof area of 2,200 square feet.

Uniform Plumbing Code

Answer: B
Code response: See Table D-1 which shows that based on a 4-inch per hour rainfall, a 3-inch leader will serve a maximum projected roof area of 2,200 square feet.

12-21 Given: The required leader size to drain a 7,550 square foot portion of a roof is 5 inches. The area rainfall is 4 inches per hour. The plumber discovers that the leader will need a 10-foot horizontal offset of ¼-inch slope per foot at the second floor. Which of the following statements is most nearly correct, as mandated by Code?

(A) Since the leader is sized to drain up to 8,650 square feet, no change in pipe size is required

(B) Only the horizontal section of piping needs to be increased in size

(C) The horizontal and vertical sections to the building drain need to be increased in size

(D) The leader piping from the building drain to the roof drain connection needs to be increased in size

---

Standard Plumbing Code

Answer: C
Code response: See Tables 1506.1 and 1506.2. The horizontal and vertical section to the building drain needs to be increased to horizontal storm drain size of 6 inches.

Uniform Plumbing Code

Answer: C
Code response: See Tables D-1 and D-2. The horizontal and vertical section to the building drain needs to be increased to horizontal rain water piping size of 6 inches.

12-22 You are installing a cast iron bell-and-spigot rain water system. According to Code, after caulking the lead, the finished joint cannot extend more than _____ inch below the rim of the hub.

(A) 1/32

(B) 1/16

(C) 1/8

(D) 3/16

Standard Plumbing Code

Answer: C
Code response: Caulked joints shall not extend more than 1/8-inch below the rim of hub.

Uniform Plumbing Code

Answer: C
Code response: After caulking, the finished joint shall not extend more than 1/8-inch below the rim of hub.

12-23 You are ready to call for inspection on a cast iron bell-and-spigot rain water system. Following the dictates of the Code, which of the following statements is most nearly correct?

(A) Code permits the painting of joints with varnish after the work is inspected and approved

(B) Code permits the painting of joints after the system has been tested

(C) Code does **not** permit the painting of lead joints before or after testing.

(D) The inspector may permit approved coatings to be applied to lead joints, **only** to seal small leaks

---

Standard Plumbing Code

Answer: A
Code response: No paint, varnish or other coatings shall be permitted on the jointing material until after the joint has been tested and approved.

Uniform Plumbing Code

Answer: A
Code response: No paint, varnish or other coatings shall be permitted on the jointing material until after the joint has been tested and approved.

Chapter 13

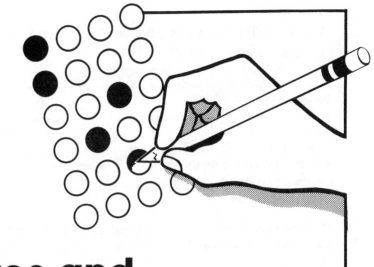

Plumbing Fixtures and Special Plumbing Fixtures

Plumbing Fixtures

The plumbing industry has, over the past hundred years, developed well-accepted standards for plumbing fixtures. These standards affect the design of all fixtures in use today. The portion of the exam that covers plumbing fixtures tests your understanding of the standards that apply to fixture design and installation.

Plumbing fixtures are the end of the potable water supply system and the beginning of the sewage system. Liquid waste flows into the fixture and then out into the drainage system. The most commonly-used fixtures are water closets, bidets, urinals, bathtubs, shower baths, kitchen sinks, laundry trays, dishwashers, and garbage disposals.

The code requires that plumbing fixtures be made of high-quality materials. They must have surfaces that are smooth and nonabsorbent, free of defects and concealed fouling surfaces. The design must allow easy cleaning of all surfaces.

Fixtures designed to meet nationally-recognized standards are commonly made of enameled cast iron, enameled pressed steel, vitreous china, stainless steel or plastic. Plastic plumbing fixtures are becoming more popular. Fixtures constructed of pervious materials (anything that lets water pass through) are not acceptable under the code.

Code governs the connection of fixtures to the drainage system, material requirements for fixture trim and installation, sizes and types of traps, and sizes and use of continuous wastes and access panels. It also sets regulations for securing fixtures to the wall or floor and for necessary clearances for their intended use and maintenance.

Bathroom plumbing fixtures must be located in adequately lighted and ventilated rooms. If natural ventilation from a window is not available, a fan and duct are required by code.

Special Plumbing Fixtures

The most common special plumbing fixtures are fixtures that don't require the standard connection to a waste and vent system. Many of these fixtures have a drain only. Others have both a drain and water supply connection. Special plumbing fixtures include coffee urns, bar sinks, floor drains, walk-in cooler drains, glass sinks, and drinking fountains.

Special plumbing fixtures with special installation requirements are common in clinics, doctors' offices, nursing homes, hospitals, prisons and the like. Some of the more common special fixtures are bedpan hoppers, bedpan washers and sterilizers, surgeons' scrub-up sinks, and perineal baths. You won't find many special plumbing fixtures in private homes, but they're routine work for plumbers who handle commercial jobs.

13-1 According to the Code, the overflow pipe from a fixture shall be connected on the:

(A)   crown vent of a tubular P-trap

(B)   crown weir of a one-piece trap

(C)   house or inlet side of the fixture trap

(D)   outlet side of the fixture trap

Standard Plumbing Code

Uniform Plumbing Code

Answer: C

Code response: The overflow pipe from a fixture shall be connected on the house or inlet side of the fixture trap.

Answer: C

Code response: The overflow pipe from a fixture shall be connected on the house or inlet side of the fixture trap.

13-2 Plumbing fixtures shall be constructed of approved materials. Which of the following materials are not acceptable by the Code?

(A)   china

(B)   cultured marble

(C)   materials that are impervious

(D)   materials that are pervious

Standard Plumbing Code

Uniform Plumbing Code

Answer: D

Code response: Plumbing fixtures shall be constructed from approved materials, and have impervious surfaces.

Answer: D

Code response: Plumbing fixtures shall have impervious surfaces.

13-3 According to the Code, plumbing fixtures shall be free from:

(A)   smooth surfaces

(B)   chipped surfaces

(C)   impervious surfaces

(D)   less dense surfaces

Standard Plumbing Code

Uniform Plumbing Code

Answer: B

Code response: Plumbing fixtures shall be free from defects.

Answer: B

Code response: Plumbing fixtures shall be free from defects.

13-4 According to the Code, built-in bathtubs with overhead showers shall have:

(A)   a watertight joint between the tub and wall

(B)   an 8-inch shower arm

(C)   a minimum of 4-foot tile walls

(D)   either a shower door or curtain

Standard Plumbing Code

Uniform Plumbing Code

Answer: A

Code response: The joint where a fixture comes in contact with a wall shall be watertight.

Answer: A

Code response: Where a fixture comes in contact with a wall, the joint shall be made watertight.

Q 13-5 According to the Code, using Figure 13-1, the distance "X" for a water closet must be a minimum of _____ inches.

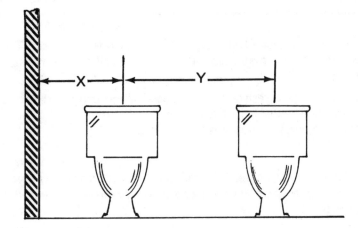

Figure 13-1

(A) 12
(B) 15

(C) 24
(D) 30

Standard Plumbing Code

Uniform Plumbing Code

Answer: B
Code response: No water closet shall be set closer than 15 inches from its center to any side wall or partition.

Answer: B
Code response: No water closet shall be set closer than 15 inches from its center to any side wall or obstruction.

Q 13-6 Again, using Figure 13-1, according to the Code, when water closets are set in a battery installation, distance "Y" shall be a minimum of _____ inches.

(A) 30
(B) 24

(C) 15
(D) 12

Standard Plumbing Code

Uniform Plumbing Code

Answer: A
Code response: No water closet shall be set closer than 30 inches center-to-center.

Answer: A
Code response: No water closet shall be set closer than 30 inches center-to-center.

13-7 All porcelain enamel surfaces on plumbing fixtures shall be:

(A) shatter resistant
(B) acid resistant

(C) heat resistant
(D) freeze resistant

Standard Plumbing Code

Uniform Plumbing Code

Answer: B
Code response: Refer to American National Standards Institute, Inc. (ANSI) materials for plumbing installations.

Answer: B
Code response: The enamel surfaces on fixtures shall be acid resistant.

Note: Although the term "acid resistant" is not used specifically in the text of the Standard Plumbing Code, American National Standards Institute, Inc. (ANSI) requires that plumbing fixtures have an acid-resistant finish. The correct answer is "B".

13-8 A 30-gallon electric water heater is installed under a staircase in a two-story residence. The relief valve drain line, according to Code, must extend to the exterior of the building and terminate a minimum of ____ inches above grade.

(A) 6
(B) 8

(C) 10
(D) 12

Standard Plumbing Code

Uniform Plumbing Code

Answer: A
Code response: The discharge from the relief valve shall be piped to the outside of the building and terminate no less than 6 inches above grade.

Answer: A
Code response: Relief valves located inside a building shall drain to the outside of the building with the end of the pipe a minimum of 6 inches above ground and pointing downward.

13-9 A 40-gallon water heater requires a combination temperature and pressure relief valve. The valve drain outlet is ½-inch. According to Code the drain line, then, should be ____ inch.

(A) 1/4
(B) 3/8

(C) 1/2
(D) 5/8

Standard Plumbing Code

Uniform Plumbing Code

Answer: C
Code response: The discharge from the relief valve shall be piped full-size, separately, to the outside of the building.

Answer: C
Code response: Relief valves located inside a building shall be provided with a full-size drain and piped to the outside of the building.

13-10 The one of the following which is prohibited by Code for restaurant use is the:

(A) siphon jet water closet
(B) washout-type water closet

(C) reverse trap water closet
(D) siphon action water closet

Standard Plumbing Code

Uniform Plumbing Code

Answer: B
Code response: A washout water closet is a prohibited fixture.

Answer: B
Code response: A washout water closet is a prohibited fixture.

13-11 According to the Code, a commercial dishwashing machine must be:

(A) directly connected to building sanitary drain
(B) indirectly connected to building sanitary drain

(C) directly connected to building greasy waste line
(D) indirectly connected to building greasy waste line

___

Standard Plumbing Code

Uniform Plumbing Code

Answer: D
Code response: A dishwasher **shall not be directly** connected to a drainage system.

Answer: D
Code response: No dishwashing machine shall be directly connected to a drainage system.

13-12 According to the Code, a commercial water closet (flush valve type) must have a minimum ____-inch(es) water supply pipe connection.

(A) 1/2
(B) 3/4

(C) 1
(D) 1¼

___

Standard Plumbing Code

Uniform Plumbing Code

Answer: C
Code response: The minimum size of a water closet supply pipe (flush valve type) shall be 1-inch.

Answer: C
Code response: The minimum size pipe connection for a water closet (flushometer type) shall be 1-inch.

13-13 A water heater must have a shut-off valve in the cold water supply. According to Code, the valve location should not exceed ____ feet from the water heater.

(A) 3
(B) 4

(C) 5
(D) 6

___

Standard Plumbing Code

Uniform Plumbing Code

Answer: A
Code response: A water heater shut-off valve shall be accessible and within 3 feet of the heater or tank.

Answer: A
Code response: A fullway (gate) valve shall be installed in the cold water supply pipe to each water heater at or near the water heater.

Note: Three feet is considered by most Codes to be a reasonable distance from a water heater to locate the control valve. Answer "A" is correct.

13-14 Many Codes consider floor drains as plumbing fixtures. According to Code, which of the following statements is most nearly correct?

(A) It is not necessary to install floor drains in public toilet rooms
(B) The Code mandates the installation of floor drains in public toilet rooms

(C) The drain inlet must be located in full view
(D) Floor drains must be located where strainers may not cause a hazard

___

Standard Plumbing Code

Uniform Plumbing Code

Answer: C
Code response: Floor drain inlets shall be so located that they are, at all times, in full view.

Answer: C
Code response: Floor drain inlets shall be so located that they are, at all times, in full view.

13-15 Bathtubs are generally installed with three sides recessed into the walls of a bathroom. According to Code, _____ must be provided.

(A) an access panel
(B) a waste and overflow of the straight-through type

(C) a waste opening with a strainer
(D) a waste that will close tight

Standard Plumbing Code

Uniform Plumbing Code

Answer: A
Code response: Fixtures having concealed slip-joint connections shall be provided with an access panel.

Answer: A
Code response: Fixtures having concealed slip-joint connections shall be provided with an access panel.

13-16 Given: You are replacing a galvanized steel pipe lavatory drain with copper pipe. The connection to the cast-iron stack must be made with a solder bushing. Code requires that the minimum weight of the solder bushing be _____ ounces.

(A) 4
(B) 6

(C) 8
(D) 10

Standard Plumbing Code

Uniform Plumbing Code

Answer: B
Code response: 1¼-inch soldering bushings where permitted shall have a minimum weight of 6 ounces.

Answer: B
Code response: 1¼-inch soldering bushings when used shall have a minimum weight of not less than 6 ounces.

13-17 According to Code, when installing wall-hung water closets:

(A) the bottom of the water closet flange must be set on an approved firm base
(B) the water closet bowl must be of the elongated type

(C) the water closet must be securely bolted to an approved carrier fitting
(D) the closet bend must be cut off so as to present a smooth surface

---

Standard Plumbing Code

Uniform Plumbing Code

Answer: C
Code response: Wall-hung water closet bowls shall be rigidly supported by a concealed metal supporting member.

Answer: C
Code response: Wall-mounted water closet fixtures shall be securely bolted to an approved carrier fitting.

13-18 Some of the more common wall-supported plumbing fixtures are water closets, urinals and lavatories. The Code requires that such fixtures be rigidly supported so that:

(A) no strain is transmitted to the connections
(B) they will not likely loosen and fall

(C) they may serve their intended purpose
(D) each fixture is readily accessible for cleaning

Standard Plumbing Code

Answer: A
Code response: Wall-hung water closet bowls shall be rigidly supported by a concealed metal supporting member so that no strain is transmitted to the closet connection.

Uniform Plumbing Code

Answer: A
Code response: Wall-hung fixtures shall be rigidly supported by metal supporting members so that no strain is transmitted to the connections.

Note: The Standard Plumbing Code addresses only wall-hung water closets. The Uniform Plumbing Code addresses all wall-hung fixtures. Answer "A" is correct.

13-19 When roughing-in the waste and water piping, according to Code, consideration should be given so that when fixtures are installed later:

(A) all pipes are afforded easy access

(B) the vertical distance between fixture bottom and the trap weir shall not exceed minimum length

(C) all pipes from fixtures shall be run to the nearest wall
(D) they may be set at standard height from finished floor

Standard Plumbing Code

Answer: C
Code response: Where practical, all pipes from fixtures shall be run to the nearest wall.

Uniform Plumbing Code

Answer: C
Code response: Where practical, all pipes from fixtures shall be run to the nearest wall.

13-20 You are estimating the lavatory needs for an industrial building. Circular basin floor mounted fixtures are to be used. According to Code, each _____ inches is considered equivalent to 1 lavatory.

(A) 12
(B) 18

(C) 20
(D) 24

Standard Plumbing Code

Answer: B
Code response: 18 inches of a circular basin shall be considered equivalent to 1 lavatory.

Uniform Plumbing Code

Answer: B
Code response: 18 inches of a circular basin shall be considered equivalent to 1 lavatory.

13-21 Sometimes workers in industrial plants are exposed to skin contamination from irritating materials. Where this type of work condition exists, the Code requires that 1 lavatory be installed for each ___ persons.

(A) 2 (C) 4
(B) 3 (D) 5

Standard Plumbing Code | Uniform Plumbing Code

Answer: D
Code response: Where there is exposure to skin contamination, provide 1 lavatory for each 5 persons.

Answer: D
Code response: Where there is exposure to skin contamination, provide 1 lavatory for each 5 persons.

13-22 A construction job employing a maximum of 90 workers must provide, according to Code, a minimum of ___ temporary toilet facilities (chemical toilets) for their use.

(A) 1 (C) 3
(B) 2 (D) 4

Standard Plumbing Code | Uniform Plumbing Code

Answer: C
Code response: Temporary facilities for use by workmen during building construction, shall consist of at least 1 water closet and 1 urinal for each 30 workmen.

Answer: C
Code response: Temporary workingmen facilities require 1 water closet and 1 urinal for each 30 workmen.

13-23 You are estimating the urinal and water closet needs for a warehouse. The facilities are centrally located. The owner states that maximum male employment will not exceed 70. To comply with Code, the owner must provide:

(A) no urinals and 4 water closets (C) 2 urinals and 2 water closets
(B) 1 urinal and 3 water closets (D) 3 urinals and 1 water closet

Standard Plumbing Code | Uniform Plumbing Code

Answer: A
Code response: Manufacturing warehouses, workshops, loft buildings, foundries, and similar establishments are not required to have urinals for employee use.

Answer: A
Code response: Industrial warehouses, workshops, foundries and similar establishments are not required to have urinals for employee use.

Note: For **local** exams, check **local** Code requirements. Many local Codes require urinals to be installed in men's toilet rooms where more than 5 males use the facilities.

13-24 Code mandates that flushing tanks used to flush urinals be:

(A)   manual in operation

(B)   limited to a minimum of 3 urinals

(C)   limited to a maximum of 6 urinals

(D)   automatic in operation

Standard Plumbing Code

Answer: D
Code response: Tanks flushing more than 1 urinal shall be automatic in operation.

Uniform Plumbing Code

Answer: D
Code response: Tanks flushing more than 1 urinal shall be automatic in operation.

13-25 Using Figure 13-2, Code requires that distance "X" for a wall-hung urinal be a minimum of ____ inches.

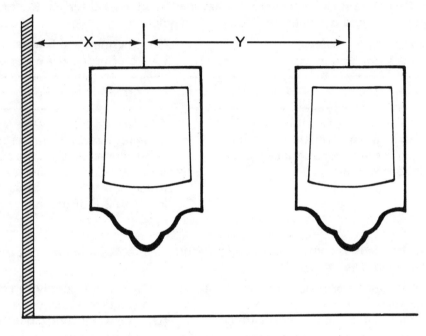

Figure 13-2

(A)   12

(B)   16

(C)   18

(D)   24

Standard Plumbing Code

Answer: A
Code response: No urinal shall be set closer than 12 inches from its center to any side wall or partition.

Uniform Plumbing Code

Answer: A
Code response: No urinal shall be set closer than 12 inches from its center to any side wall or partition.

Q 13-26 Again, using Figure 13-2, when urinals are set in a battery installation, Code would require that distance "Y" be a minimum of _____ inches.

(A) 12

(B) 16

(C) 18

(D) 24

Standard Plumbing Code	Uniform Plumbing Code
Answer: D Code response: No urinal shall be set closer than 24 inches center-to-center with an adjacent urinal fixture.	**Answer: D** Code response: No urinal shall be set closer than 24 inches center-to-center.

Q 13-27 You are estimating the urinal and water closet needs for common toilet facilities in a commercial building for multiple tenant use. The architect has specified the need for 12 water closets. In lieu of the 12 water closets, you could comply with Code by installing:

(A) 1 urinal and 11 water closets

(B) 2 urinals and 10 water closets

(C) 3 urinals and 9 water closets

(D) 5 urinals and 7 water closets

Standard Plumbing Code	Uniform Plumbing Code
Answer: C Code response: One urinal required for each 3 required water closets in all men's toilet rooms when accessible to the public.	**Answer: C** Code response: Whenever urinals are provided, 1 water closet less than the number may be provided for each urinal installed, except the number of water closets in such cases shall not be reduced to less than 2/3 of the minimum specified.

Q 13-28 A church baptistry is considered a special fixture. To meet Code requirements when installing the water piping for a baptistry:

(A) the baptistry must be provided with both hot and cold water

(B) the water supply must be provided with an air gap

(C) the water supply can be no less than that required for similar fixtures

(D) the water piping has to be insulated for protection

Standard Plumbing Code	Uniform Plumbing Code
Answer: B Code response: Baptistries when provided with water supplies shall be protected from back-siphonage.	**Answer: B** Code response: Special fixtures when provided with water supplies shall be protected from back-siphonage.

13-29 An industrial plant is designed with multiple wash sinks of the straight-line type in lieu of single lavatories. The need is equal to 6 lavatories. According to Code, the length of the wash sink can be no less than ____ inches.

(A) 48
(B) 72

(C) 108
(D) 144

Standard Plumbing Code

Uniform Plumbing Code

Answer: D
Code response: 24 lineal inches of wash sink shall be considered equivalent to 1 lavatory.

Answer: D
Code response: 24 lineal inches of wash sink shall be considered equivalent to 1 lavatory.

13-30 Most flushometer valves for urinals are manually controlled. According to Code, if water pressure is adequate, a single flushometer valve may be used to flush ____ urinal(s).

(A) 1
(B) 2

(C) 3
(D) 4

Standard Plumbing Code

Uniform Plumbing Code

Answer: A
Code response: No flushometer valve shall be used to flush more than 1 urinal.

Answer: A
Code response: No manually controlled flushometer valve shall be used to flush more than 1 urinal.

13-31 Code prohibits the installation of a drinking fountain in:

(A) a hallway with heavy pedestrian traffic
(B) a toilet room that is well-lighted and ventilated

(C) the entrance way to a public building
(D) an exit reserved for emergency use

Standard Plumbing Code

Uniform Plumbing Code

Answer: B
Code response: Drinking fountains shall not be installed in toilet rooms.

Answer: B
Code response: Drinking fountains shall not be installed in toilet rooms.

13-32 The Code prohibits the installation of any water closet which might permit:

(A) an unreasonable amount of noise when flushed

(B) a whirling motion in the water as it is discharged

(C) the emptying of 2/3 of the bowl water with each flush
(D) an unventilated space

Standard Plumbing Code

Uniform Plumbing Code

Answer: D
Code response: Water closets having an unventilated space are prohibited.

Answer: D
Code response: Water closets having an unventilated space are prohibited.

13-33 According to Code, shower drains must:

(A) be constructed of cast iron with chrome strainers
(B) have adequate weep holes
(C) be set level with the shower floor
(D) be no smaller than 1½-inches in diameter

Standard Plumbing Code | Uniform Plumbing Code

Answer: B
Code response: Shower drain shall have an approved weep hole device system to insure constant drainage of water from the shower pan to sanitary system.

Answer: B
Code response: The subdrain (shower drain) shall have weep holes into the waste line.

13-34 To make a watertight connection between the shower drain and a tiled shower compartment, the Code requires:

(A) that tile be properly grouted where it joins the shower strainer
(B) no less than approved linings be installed
(C) that shower drains be equipped with a clamping ring
(D) that floors under shower compartments be smooth

Standard Plumbing Code | Uniform Plumbing Code

Answer: C
Code response: Shower drains shall be so constructed with a clamping device that the pan may be securely fastened to the shower drain, thereby making a watertight joint.

Answer: C
Code response: Each such subdrain (shower drain) shall be equipped with a clamping ring to make a tight connection between the lining and the drain.

Questions 13-35 through 13-40 will relate to "Requirements for the Physically Handicapped." Oddly enough, neither Code addresses these requirements. However, because of the stringent Federal guidelines legislated in recent years, the American National Standards Institute, ANSI A 117.1-1961 (R1971) has developed regulations implementing those Federal guidelines. Local Codes are charged with enforcing them.

All required plumbing facilities (other than single family or apartments) must now provide accessibility for the physically handicapped. Although questions on this topic may or may not be included for state exams, there is a strong possibility they will be included in local exams.

This is an important part of plumbing work. As a plumbing tradesman, you need to be as informed as possible on such new requirements so that you don't have to re-do work.

13-35 A public toilet room having 6 regular water closets, according to ANSI, must provide ____ water closet(s) for the physically handicapped.

(A) 1
(B) 2
(C) 3
(D) 4

American National Standards Institute

Answer: A
ANSI response: Toilet rooms shall have at least 1 toilet accessible and usable by the physically handicapped.

Q 13-36 When installing a wall-hung lavatory for the physically handicapped, the maximum height from the finished floor to the top of the lavatory must be _____ inches.

(A) 31　　　　　　　　　　　　　　　　(C) 33
(B) 32　　　　　　　　　　　　　　　　(D) 34

American National Standards Institute

Answer: D

ANSI response: Toilet rooms shall have at least 1 lavatory installed that is usable by individuals in wheelchairs. The maximum height for such lavatory is 34 inches.

Q 13-37 The Code mandates that a water closet for conventional use be set no closer than 15 inches from the center of the bowl to any wall or partition. However, ANSI mandates that water closets for the physically handicapped have this minimum distance to be _____ inches.

(A) 16　　　　　　　　　　　　　　　　(C) 18
(B) 17　　　　　　　　　　　　　　　　(D) 20

American National Standards Institute

Answer: C

ANSI response: One water closet shall be set no closer than 18 inches to any wall or partition.

Q 13-38 The standard height from the finished floor to the top of the bowl for a conventional water closet is approximately 14½ inches. According to ANSI, the height of a handicapped seat shall be most nearly _____ inches.

(A) 16　　　　　　　　　　　　　　　　(C) 21
(B) 20　　　　　　　　　　　　　　　　(D) 22

American National Standards Institute

Answer: B

ANSI response: Toilet rooms shall have a water closet with the seat 20 inches from the floor.

Q 13-39 Toilet rooms for men usually have wall-mounted urinals. The opening of a standard mounted urinal from the floor is approximately 24 inches. According to ANSI, the opening of a urinal for the handicapped must be no higher than _____ inches from the floor.

(A) 19　　　　　　　　　　　　　　　　(C) 21
(B) 20　　　　　　　　　　　　　　　　(D) 22

American National Standards Institute

Answer: A

ANSI response: Toilet rooms for men shall have wall-mounted urinals with the opening of the basin 19 inches from the floor.

Q 13-40 Water fountains shall be accessible to and usable by the physically handicapped. According to ANSI, wall-mounted drinking fountains must have a maximum height of _____ inches from finished floor to top.

(A) 30

(B) 32

(C) 34

(D) 36

American National Standards Institute

Answer: D

ANSI response: Wall-mounted drinking fountains can serve the able-bodied and the physically disabled equally well when mounted with the basin 36 inches from the floor.

Part Two
Gas Systems

- Gas Regulations and Definitions
- Gas Piping—Materials and Installation
- Gas Appliances, Vents and Flues

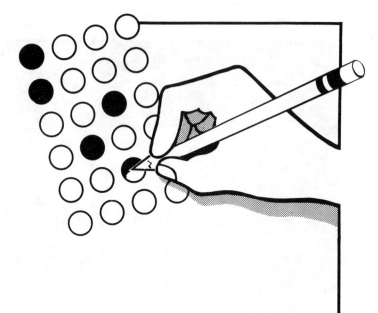

Chapter 14

Gas Regulations and Definitions

Natural gas has been used in China for over 3,000 years. The Chinese are credited with being the first to use this fuel for heating and cooking.

Today, gas piping is an important part of plumbers' and gas installers' work. Gas is used for cooking, refrigeration and heating, as well as many industrial applications.

Plumbers and gas installers need to be familiar with three types of gas: *natural gas, manufactured gas* and *liquified petroleum gas* (better known as LPG). Only manufactured gas is considered poisonous. But all three can explode or cause suffocation. Any time you're installing gas piping, be aware that there's a potential hazard to life and property. That's why there are gas codes and why the codes are strictly enforced.

General Regulations

Gas codes regulate who can install gas piping, and provide rules for the installation, operation and maintenance of gas piping and gas equipment on consumers' premises. The gas code refers to the manufacturers' recommendations, gas suppliers' regulations, and other standards.

The questions in this chapter will test your understanding of the applicable regulations for gas systems.

Definitions

The second part of this chapter covers definitions of gas terms. This is an important part of following the gas code. You can't interpret or apply regulations if you don't understand the terms used. You'll have several questions on both gas regulations and code definitions on your exam.

14-1 The Code requires that location of the gas meter be acceptable to the:

(A) consumer
(B) gas supplier

(C) plumbing contractor
(D) plumbing official

Standard Gas Code

Answer: B

Code response: The gas meter location shall be acceptable to the gas company.

14-2 According to Code, multiple-meter gas piping installations which serve separate gas systems must be:

(A) accessible
(B) a minimum of 1 inch in diameter

(C) readily identified
(D) approved by the gas supplier

Standard Gas Code

Answer: C

Code response: Piping from multiple-meter installations shall be plainly marked by the installer so that each system can be readily identified at meter locations.

14-3 The Code may require that your first responsibility, before proceeding with the installation of a gas piping system, is to:

(A) check the job site
(B) prepare a piping diagram

(C) secure a gas permit
(D) prepare a material take-off for the job

Standard Gas Code

Answer: B

Code response: Before proceeding with the installation of a gas piping system, a piping sketch or plan may be required.

14-4 When it becomes necessary to connect any additional gas appliance to an existing gas system, the Code requires verification that:

(A) existing gas piping is properly supported
(B) existing gas piping has adequate capacity

(C) a master plumber supervises the work
(D) existing gas piping is in good physical condition

Standard Gas Code

Answer: B

Code response: The capacity of the existing gas piping shall be verified as being adequate when an additional appliance is to be served.

14-5 Code requires that gas piping which serves more than one building on the same premises and which utilitizes 2 or more gas meters, must:

(A) not be interconnected on the outlet side of the meters

(B) have a separate shut-off valve near each meter

(C) have ground joint unions at each meter location

(D) be protected from corrosion

Standard Gas Code

Answer: A

Code response: When 2 or more meters are installed on the same premises but supply separate consumers, the gas piping shall not be interconnected on the outlet side of the meters.

14-6 Given: A bedroom is being added to an existing residence. The gas meter must be relocated. According to Code, the moving of the gas meter must be done by:

(A) the plumbing contractor's employees

(B) a certified gas contractor's employees

(C) gas company employees

(D) a specialty contractor's employees certified in gas piping

Standard Gas Code

Answer: C

Code response: The moving of gas meters shall be done by gas company employees.

14-7 According to Code, gas meters must be located in ventilated spaces and be:

(A) readily accessible for reading

(B) preceded by a bypass connection

(C) preceded by a house shut-off valve

(D) firmly secured to an outer building wall

Standard Gas Code

Answer: A

Code response: A meter location shall be such that the meter and connections are easily accessible in order that the meter may be read or changed.

14-8 According to Code, a permit must be obtained before a plumbing contractor's employee can:

(A) install a gas meter

(B) install a conversion burner

(C) extend a gas supplier's service line

(D) light a water heater pilot light

Standard Gas Code

Answer: B

Code response: A person shall not install a gas conversion burner without first obtaining a permit.

14-9 A gas company's employees are installing a new service line and meter for a dry cleaning establishment. According to Code:

(A) the work must be inspected by the plumbing official

(B) the gas company shall first obtain a permit

(C) the gas company does not need a permit

(D) the installation of the service line and meter shall be in accordance with approved plans

Standard Gas Code

Answer: C

Code response: The gas company shall not be required to obtain permits to set meters, or to extend its service lines.

14-10 The Gas Code governs the installation of:

(A) a gas company service line from private line to gas meter location

(B) a gas company branch line from main to within 6 feet of private property line

(C) gas piping for commercial consumers' use

(D) a public utility gas distribution system

Standard Gas Code

Answer: C

Code response: The provisions of this Code shall govern the installation of consumers' gas.

14-11 Code requires that a certified gas or plumbing contractor handle repairs, alterations, relocations or other work on any portion of consumers' gas piping containing:

(A) measured gas

(B) manufactured gas

(C) natural gas

(D) unmeasured gas

Standard Gas Code

Answer: D

Code response: Repairs, alterations, relocations or other work on any portion of consumers' gas piping containing unmeasured gas shall be done by a qualified installing agency authorized to do such work.

14-12 After completion of repairs on a consumer's gas piping containing unmeasured gas, the Code requires that the installer:

(A) call for an inspection

(B) check the piping for any leaks

(C) notify the gas company

(D) replace all existing pipe hangers

Standard Gas Code

Answer: C

Code response: When repairs are completed on a consumer's gas piping containing unmeasured gas, the installer shall notify the gas company.

Q 14-13 According to Code, when the Administrative Authority approves the use of plastic pipe or tubing in a natural gas piping system:

(A) such piping may be used only outside, and above ground

(B) such piping may be used only outside, and underground

(C) there are no restrictions as to its use

(D) such piping must be at least one pipe size larger than metal pipe

Standard Gas Code

Answer: B

Code response: When approved by the Administrative Authority, plastic pipe or tubing may be used for outside natural gas piping underground only.

Q 14-14 A gas company, at times, approves the use of seamless copper tubing which is made in several weights, called types. According to Code, the type most likely to be accepted by a gas company for interior gas piping is:

(A) L

(B) M

(C) D

(D) A

Standard Gas Code

Answer: A

Code response: When acceptable to the gas company, seamless copper tubing shall be of standard type K or L.

Q 14-15 Some gases are considered to be corrosive. Where such gases are used, the Code mandates that ____ can **not** be used for consumers' gas piping.

(A) wrought-iron pipe

(B) wrought-steel pipe

(C) brass pipe in iron pipe sizes

(D) galvanized steel pipe

Standard Gas Code

Answer: C

Code response: Consumers' gas piping of copper or brass pipe in iron pipe sizes may be used with gases **not corrosive** to such materials.

Q 14-16 Code mandates that all plastic pipe approved for use with natural gas has to:

(A) bear the manufacturer's address

(B) have the ANSI specifications stamped thereon

(C) be polyvinyl chloride, type grade 1

(D) indicate resins used in manufacturing

Standard Gas Code

Answer: D

Code response: All plastic pipe approved for natural gas shall bear the production code number to indicate resins used in manufacturing.

Q **14-17** According to Code, nonferrous pipe used in a gas piping system may be soldered. The solder acceptable for this type work is known as "hard solder" and is composed of:

(A) tin and lead

(B) zinc and silver

(C) copper and zinc

(D) tin, lead and zinc

Standard Gas Code

Answer: C

Code response: Nonferrous pipe may also be soldered using material having a melting point in excess of 1,000 degrees F.

Note: Melting points of solders that exceed 618.8 degrees F. are considered as "hard solder". Hard solder used for soldering copper is generally composed of 50 percent copper and 50 percent zinc.

Q **14-18** Copper tubing used for gas piping, according to Code, may have joints made with:

(A) compression type gas tubing fittings

(B) flared gas tubing fittings

(C) flanged type gas tubing fittings

(D) swedge type gas tubing fittings

Standard Gas Code

Answer: B

Code response: Copper tubing used in a gas piping system shall have joints made with flared gas tubing fittings, or be soldered, or brazed.

Q **14-19** Solder used for making joints in a copper tubing gas piping system, according to Code, should have a melting point in excess of _____ degrees F.

(A) 650

(B) 715

(C) 880

(D) 1,000

Standard Gas Code

Answer: D

Code response: Solder used to make tubing joints shall have a melting point in excess of 1,000 degrees F.

Q **14-20** When testing out a gas system for inspection, you discover a pin hole in an elbow. You seal the hole with a substance called "crack-stick." According to Code:

(A) this method is permitted for small holes

(B) the hole may be soldered

(C) defective fittings shall not be repaired

(D) defective fittings may be properly repaired

Standard Gas Code

Answer: C

Code response: Defects in gas piping or fittings shall not be repaired.

Q 14-21 According to Code, gas piping and fittings must be clear and free from cutting burrs and defects in structure or threading, and they also must be:

(A) properly supported from the ceiling

(B) thoroughly brushed and scale blown

(C) thoroughly brushed and coated with a noncorrosive paint

(D) protected against excessive temperatures

Standard Gas Code

Answer: B

Code response: Gas piping and fittings shall be clear and free from cutting burrs and defects in structure or threading and shall be thoroughly brushed and scale blown.

Q 14-22 According to Code, the threads of metallic piping used in gas systems must comply with the ___ for pipe threads.

(A) ANSI B2.1-1968

(B) ASTM B88-1981

(C) NFPA 96-1980

(D) AGA 56-1982

Standard Gas Code

Answer: A

Code response: Pipe and fitting threads shall comply with the ANSI Standard for Pipe Threads, B2.1-1968.

Q 14-23 The Code makes it clear that when installing polyethylene Type III plastic pipe in contact with material exerting a corrosive action:

(A) it must be coated with a corrosion resistant material

(B) it must be wrapped for protection against exterior corrosion

(C) the joints must be of the heat-fusion type

(D) the provisions of this Code do not apply

Standard Gas Code

Answer: D

Code response: When in contact with material exerting a corrosive action, **metallic piping and fittings** shall be coated with a corrosion resistant material.

Q 14-24 The approximate length of threads for a 2-inch wrought-steel pipe, according to Code, would be most nearly ___ inch(es).

(A) 3/4

(B) 7/8

(C) 1

(D) 1¼

Standard Gas Code

Answer: C

Code response: The approximate length of threads for a 2-inch threadable pipe is 1 inch.

Note: See **Standard Gas Code**, "Specifications for Threaded Pipe," Table 4.

Q 14-25 According to Code, in a wrought-iron gas piping system, the fittings to connect piping 4 inches and larger:

(A) must be shop welded

(B) may be cast-iron

(C) should be of the flange type

(D) must be compression type mechanical joints

Standard Gas Code

Answer: B

Code response: Cast-iron fittings in sizes 4 inches and larger may be used to connect steel and wrought-iron pipe.

Q 14-26 According to Code, PVC-2110 plastic pipe 1-inch and larger (when acceptable for use as a gas service line by the Administrative Authority and gas company):

(A) cannot be threaded

(B) may be threaded

(C) should have threads that comply with ANSI requirements

(D) must be properly supported

Standard Gas Code

Answer: A

Code response: Plastic pipe or tubing shall not be threaded.

Q 14-27 Solvent cement or heat-fusion joints are permitted by Code for the joining of:

(A) polyethylene and polyvinyl chloride plastics

(B) PE and PVC plastics

(C) polyvinyl chloride and PVC plastics

(D) polyvinyl chloride and PE plastics

Standard Gas Code

Answer: C

Code response: Solvent cement or heat-fusion joints shall not be made between different kinds of plastic.

Note: Since PVC is the abbreviation for polyvinyl chloride plastic, clearly, it is permissible to fuse these like materials. Answer "C" is correct.

Q 14-28 In a gas system, it sometimes becomes necessary to connect plastic to metallic piping. The minimum horizontal length of metallic piping underground required by Code at the end of any plastic piping installed, is _____ inches.

(A) 10

(B) 12

(C) 14

(D) 16

Standard Gas Code

Answer: B

Code response: When connecting plastic to a metallic riser, there shall be a minimum 12-inch horizontal length of metallic piping underground at the end of any plastic piping installed.

Q 14-29 When it becomes necessary to connect plastic pipe to metallic pipe in a gas system, Code requires that the connection be made:

(A) outside of the building, underground

(B) outside of the building, above ground

(C) with compression type mechanical joints only

(D) with wrought-iron pipe and fittings only

Standard Gas Code

Answer: A

Code response: Connections between plastic and metallic pipe shall be made outside of the building and underground.

Q 14-30 Code mandates that when solvent cement joints are made in accordance with qualified procedures, they:

(A) need not be inspected before covering

(B) need not be tested to prove their strength

(C) must be at least as strong as the pipe being joined

(D) will not be as strong as the pipes being joined

Standard Gas Code

Answer: C

Code response: Solvent cement joints shall be made in accordance with qualified procedures which have been established and proven by test to produce gas-tight joints at least as strong as the pipe being joined.

Q 14-31 The Gas Code defines an adjustable device for varying the size of the primary air inlet of a gas appliance as an:

(A) air mixer

(B) air shutter

(C) baffle

(D) draft regulator

Standard Gas Code

Answer: B

Code response: An air shutter is an adjustable device for varying the size of the primary air inlet.

Q 14-32 Consumers' gas piping is defined by Code as "all gas piping and fittings extending from:

(A) the gas company main to the house meter."

(B) a point 6 feet from the building outside wall to the nearest appliance."

(C) the outlet side of the gas meter to all outlets (excluding the meter)."

(D) the outlet side of the gas meter to all outlets (including the meter)."

Standard Gas Code

Answer: C

Code response: Consumers' gas piping includes all the gas piping and fittings extending from the point of delivery (meter) to the outlets.

14-33 A low point in a gas piping system which collects condensate is defined by the Code as a _____.

(A) trap   (C) drip

(B) orifice   (D) hose

Standard Gas Code

Answer: C

Code response: A drip is a container placed at a low point in a gas piping system to collect condensate and from which it may be removed.

14-34 According to Code definition, _____ is a liquid which separates from a flue gas due to a reduction in temperature.

(A) acid   (C) LP gas

(B) effluent   (D) condensation

Standard Gas Code

Answer: D

Code response: Condensation is the liquid which separates from a flue gas due to a reduction in temperature.

14-35 A pipe which conveys gas from a supply line to the appliance is defined by Code as a:

(A) branch line   (C) gas main

(B) dead end   (D) house line

Standard Gas Code

Answer: A

Code response: A branch line is the gas piping which conveys gas from a supply line to the appliance.

14-36 Air that is supplied to the flame at the point of combustion is defined by Code as:

(A) secondary air   (C) flue air

(B) primary air   (D) combustion air

Standard Gas Code

Answer: A

Code response: Secondary air is air that is externally supplied to the flame at the point of combustion.

14-37 A device that is responsive to changes in pressure or temperature on a gas supply to an appliance is defined by Code as a:

(A) thermostat   (C) limit control

(B) meter   (D) pre-mixing burner

Standard Gas Code

Answer: C

Code response: A limit control is a device responsive to changes in pressure or temperature for turning on, or shutting off, or throttling, the gas supply to an appliance.

14-38 The Code defines air which enters a draft hood or draft regulator and mixes with flue gases as:

(A) dilution air

(B) secondary air

(C) combustion air

(D) flue air

Standard Gas Code

Answer: A

Code response: Dilution air is air which enters a draft hood or draft regulator and mixes with flue gases.

14-39 When flame failure occurs on a burner or a group of burners, a device which automatically shuts off the gas is defined by Code as a:

(A) modulating orifice

(B) stop cock

(C) flame safeguard

(D) limit control

Standard Gas Code

Answer: C

Code response: A flame safeguard is a device which will automatically shut off the gas supply to a main burner or group of burners when the means of ignition of such burners becomes inoperative.

14-40 A flue collar, as defined by Code, is that portion of an appliance designed for the attachment of the:

(A) flue exhauster

(B) draft hood

(C) draft regulator

(D) flue pipe

Standard Gas Code

Answer: B

Code response: A flue collar is that portion of an appliance designed for the attachment of the draft hood.

14-41 The Code defines the products of combustion from gas appliances, excess air, and diluted air above the draft hood as:

(A) relief gases

(B) measured gases

(C) vent gases

(D) unmeasured gases

Standard Gas Code

Answer: C

Code response: Vent gases are products of combustion from gas appliances plus excess air, plus dilution air in the vent connector, above the draft hood.

Q 14-42 The air that is introduced into a burner and which mixes with the gas before it reaches the port of ports, is defined by Code as:

(A) primary air

(B) secondary air

(C) transition air

(D) measured air

Standard Gas Code

Answer: A

Code response: Primary air is air that is introduced into a burner and which mixes with the gas before it reaches the port of ports.

Q 14-43 The object placed in an appliance to change the direction of, or retard the flow of, air, air-gas mixtures, or flue gases is defined by Code as a:

(A) diversity factor

(B) baffle

(C) appliance flue

(D) draft regulator

Standard Gas Code

Answer: B

Code response: A baffle is an object placed in an appliance to change the direction of, or retard the flow of, air, air-gas mixtures, or flue gases.

Q 14-44 An opening in a device which limits the flow of gas to the burners is defined by Code as an:

(A) automatic valve

(B) orifice

(C) relief opening

(D) firing valve

Standard Gas Code

Answer: B

Code response: An orifice is the opening in a cap, spud or other device, whereby the flow of gas is limited, and through which the gas is discharged to the burner.

Q 14-45 The Code defines gas vents which are used in venting gas appliances equipped to burn <u>only</u> gas as:

(A) type BW

(B) type L

(C) type B

(D) type D

Standard Gas Code

Answer: C

Code response: Type B gas vents are Code-approved for venting listed or approved appliances equipped to burn only gas.

Q

14-46 Gas vents that are approved for use in venting gas or oil burning appliances are defined by Code as:

(A) type BW

(B) type L

(C) type D

(D) type B

Standard Gas Code

Answer: B

Code response: Type L gas vents are Code-approved for the venting of listed or approved appliances equipped to burn gas or oil.

Chapter 15

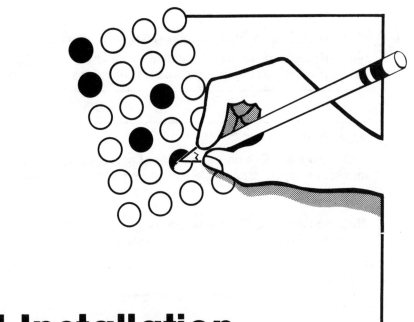

Gas Piping — Materials and Installation

The gas code spells out minimum quality and weight standards for all gas materials. These standards protect public health, welfare and safety by defining correct installation and maintenance procedures for all gas piping and appliances.

In general, materials must be free from defects. Pipe, pipe fittings and appliances must be listed or labeled by an approved listing agency. The manufacturer applies this label to certify that the product conforms with standards established by the code.

New plumbing products are developed from time to time when new materials are available, when new manufacturing methods are developed, and when construction requirements change. But not all new products are approved immediately for use under all codes. It's your job to be sure that the products you install have been approved for use in the way you plan to use them.

Without a doubt, your exam will include questions from this section of the gas code. Be familiar with gas material and installation standards to improve your chance of passing.

15-1 According to Code, gas piping has to be sized and installed to provide:

(A) sufficient gas to meet minimum demand load
(B) sufficient gas to meet maximum demand load

(C) adequate consideration to future gas demand
(D) the anticipated gas demand

Standard Gas Code

Answer: B

Code response: The hourly volume of gas required at each outlet shall be taken as not less than the maximum hourly rating, as specified by the manufacturer of the appliance.

15-2 According to Code, low pressure gas piping systems (not in excess of 0.5 lb. per square inch) may be tested for leaks by pressurizing the system with air at a pressure of _____ inches of mercury.

(A) 2
(B) 4

(C) 6
(D) 8

Standard Gas Code

Answer: C

Code response: Low pressure (not in excess of 0.5 lb. per square inch) gas piping shall withstand a pressure of at least 6 inches of mercury.

15-3 The Code states that when pressurizing low pressure gas piping for inspection (not in excess of 0.5 lb. per square inch), it must withstand a pressure of at least 3.0 PSIG for a minimum of:

(A) 10 minutes
(B) 20 minutes

(C) 30 minutes
(D) 1 hour

Standard Gas Code

Answer: A

Code response: Low pressure (not in excess of 0.5 lb. per square inch) gas piping shall withstand a pressure of at least 3.0 PSIG for a period of not less than 10 minutes without any drop in pressure.

15-4 According to Code the unthreaded portion of a 1-inch gas outlet from concealed piping should extend through the finished wall a distance of not less than _____ inch(es).

(A) ½
(B) ¾

(C) 1
(D) 1¼

Standard Gas Code

Answer: C

Code response: The unthreaded portion of gas piping outlets shall extend at least 1-inch through finished wall.

15-5 Given: A common gas supply line is to be installed in a 100-unit high-rise apartment building. Each apartment will be equipped with a gas range and water heater. According to Code, what percent of the total connected load is used to determine the gas piping size?

(A) 10.9%

(B) 14.7%

(C) 25.3%

(D) 75.8%

Standard Gas Code

Answer: B

Code response: Demand values for use in determining gas piping size in multiple dwelling units are taken from Table 3.* Percent of total connected load for 100 units equipped with gas ranges and water heaters equals 14.7 percent for the hourly heating demand required.

* Table 3 is entitled, "Demand Values for Use in Determining Gas Piping Size in Multiple Dwelling Units".

15-6 In following the Code for installing consumers' gas piping, which one of the following factors is <u>not</u> used for determining the size of a gas pipe?

(A) the length of piping and type of fittings

(B) the maximum gas consumption to be provided

(C) the specific gravity of gas

(D) the type of piping material to be installed

Standard Gas Code

Answer: D

Code response: To determine the sizes of consumers' gas piping, the following factors are considered:

1) The **length of piping** from point of delivery to the most remote outlet.

2) The total **gas demand.**

3) Table 2 based on pressure drop of 0.5 inches water column, 0.6 **specific gravity gas.**

15-7 According to Code, low pressure commercial gas piping systems may be tested for leaks by filling the system with:

(A) inert gas

(B) oxygen

(C) peppermint

(D) smoke

Standard Gas Code

Answer: A

Code response: To test gas piping for tightness, the piping shall be filled with air or **inert gas.**

Q 15-8 Using Figure 15-1 and Code requirements, the minimum distance "X" between underground gas and water piping is:

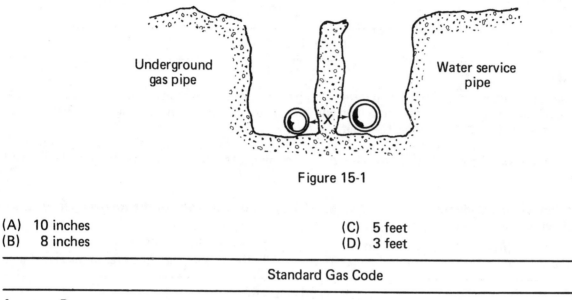

Underground
gas pipe

Water service
pipe

Figure 15-1

(A) 10 inches   (C) 5 feet
(B) 8 inches   (D) 3 feet

Standard Gas Code

Answer: B

Code response: No underground gas piping shall be installed closer than 8 inches from a water pipe.

Q 15-9 Given: You are installing a 1-inch plastic natural gas line, a 3-inch water line and a 6-inch sewer line. The distance between the property line and the building foundation is 7 feet. Due to the limited space, the minimum Code separation requirements for the 3 underground pipes cannot be met. Under these conditions the gas piping:

(A) may be installed in the same trench with the water piping
(B) may be installed in the same trench with sewer piping

(C) installation may be approved by the local Administrative Authority
(D) must be secured to the outside of the building, above grade

---
Standard Gas Code

Answer: C

Code response: Gas piping in the same ditch with water or sewer piping is prohibited, except when approved by the Administrative Authority.

15-10 **Given: You have completed the rough gas piping for a single family home containing the following gas appliances: dryer, water heater, space heater, and range. Your work was rejected by the inspector because:**

(A) the test instrument used was a manometer gauge

(B) the pressure was less than 5.0 PSIG

(C) the system was pressurized with inert gas

(D) the electrician used the gas piping in the utility room as a grounding electrode

---

Standard Gas Code

Answer: D

Code response: Gas piping shall not be used as a grounding electrode.

Note: Each of the answers "A", "B", and "C" represents a Code-accepted procedure. Answer "D" is the only one prohibited by Code.

15-11 **Given: A ¾-inch type-L copper gas line is installed in the drop ceiling of a kitchen. According to Code, the copper tubing must be supported at intervals not to exceed ____ feet.**

(A) 6

(B) 7

(C) 8

(D) 9

Standard Gas Code

Answer: A

Code response: Spacing of supports in horizontal ¾-inch gas tubing installations shall not be greater than 6 feet.

15-12 **You discover you cannot avoid the installation of a branch gas line in a solid floor. With Administrative Authority approval, the <u>one</u> of the following methods of installation which may <u>not</u> be used is:**

(A) gas piping may be installed in a casing of iron pipe with tightly sealed ends and joints

(B) gas piping may be installed in a suitably covered channel in the floor

(C) gas piping must have heat-fusion joints only

(D) gas piping may be embedded in concrete floor slabs constructed with Portland cement

---

Standard Gas Code

Answer: C

Code response: Plastic piping may not be used within or under any building or slab.

Note: Answers "A", "B" and "D" are acceptable by Code for gas piping installations with prior approval of the Administrative Authority. Answer "C", however, is **never** acceptable.

15-13 After a gas piping system is completed and checked for leakage, Code mandates that it shall:

(A) be put into service

(B) have pilots lighted

(C) be purged

(D) be connected to each appliance

Standard Gas Code

Answer: C

Code response: After the gas piping system has been checked for leakage, it shall be purged.

15-14 The gas company installs a consumer's meter on the outside wall of an accessory building. You are charged with the installation of this consumer's gas pipe from point of delivery to the primary house. Discounting cubic feet of gas per hour required, the minimum underground pipe size acceptable by Code is ____ inch(es).

(A) ¾

(B) 1

(C) 1¼

(D) 1½

Standard Gas Code

Answer: B

Code response: All consumers' gas pipe from the point of delivery (meter) to the house piping shall not be less than 1-inch in diameter.

15-15 According to Code, when installing wrought steel gas piping within a building, the minimum size for pipe outlets must be ____ inch.

(A) 1/4

(B) 3/8

(C) 1/2

(D) 3/4

Standard Gas Code

Answer: C

Code response: The minimum size of pipe outlets shall be ½-inch.

15-16 The only one of the following concealed gas piping installations not accepted by Code is piping located:

(A) in hollow partitions

(B) in an elevator shaft

(C) in solid partitions provided with a chase

(D) in solid partitions within a casing

Standard Gas Code

Answer: B

Code response: Gas piping inside any building shall not be run in or through an elevator shaft.

15-17 Assume that it becomes necessary to place a tee in a wrought-iron gas branch line for the installation of a space heater. The proper fitting to use, according to Code, in reconnecting the pipe and fittings is a:

(A) right and left coupling

(B) coupling with running threads

(C) ground joint union with the nut "center punched."

(D) union with gasproof gasket

Standard Gas Code

Answer: C

Code response: When necessary to insert fittings in pipe which has been installed in a concealed location, the pipe may be reconnected by the use of a ground joint union with the nut "center punched."

15-18 Assume that a 20-foot section of gas pipe is to be embedded in a concrete floor slab constructed of Portland cement. According to Code, the piping has to be surrounded with a minimum of _____ inch(es) of solid concrete.

(A) 1

(B) 1¼

(C) 1½

(D) 1¾

Standard Gas Code

Answer: C

Code response: Gas piping may be embedded in concrete floor slabs constructed with Portland cement. Piping shall be surrounded with a minimum of 1½ inches of solid concrete.

15-19 The approximate number of standard pipe threads per inch for a 1¼-inch galvanized steel pipe, according to Code, is:

(A) 10

(B) 11

(C) 12

(D) 13

Standard Gas Code

Answer: B

Code response: For 1¼-inch iron pipe, the approximate number of threads per inch to be cut is 11. See Table 4, "Specifications for Threaded Pipe."

15-20 When joining a piece of gas pipe with a fitting, Code requires that a joint compound be used. Which, then, of the following statements is most nearly correct?

(A) Joint compound must be heavily applied to male threads only

(B) Joint compound must be sparingly applied to male threads only

(C) Joint compound must be heavily applied to female threads only

(D) Joint compound must be sparingly applied to female threads only

Standard Gas Code

Answer: B

Code response: Joint compound shall be applied sparingly and only to the male threads of metallic joints.

15-21 According to Code, the one of the following fittings that <u>cannot</u> be used in a gas piping system is a:

(A) bushing

(B) reducing coupling

(C) flared fitting

(D) nonferrous fitting

Standard Gas Code

Answer: A

Code response: Bushings shall not be used in gas piping.

15-22 Code mandates that when compression type mechanical joints are used with plastic piping, the gasket material in the fitting be:

(A) oil and gas resistant

(B) constructed of neoprene rubber

(C) compatible with the gas distributed by the gas company

(D) asbestos, and resistant to corrosion

Standard Gas Code

Answer: C

Code response: When compression type mechanical joints are used, the gasket material in the fitting shall be compatible with the gas distributed by the gas company.

15-23 When installing gas piping (other than for dry gas), Code requires that the gas piping be properly drained of condensate. The minimum pitch per 15 feet acceptable by Code is _____-inch.

(A) 1/16

(B) 1/8

(C) 1/4

(D) 1/2

Standard Gas Code

Answer: C

Code response: Except where a dry gas is distributed, gas piping shall be properly drained with a minimum pitch of approximately ¼-inch in 15 feet.

15-24 The Code states that a drip pipe for a 1½-inch gas pipe must have a minimum diameter of _____ inch(es).

(A) 1

(B) 1¼

(C) 1½

(D) 2

Standard Gas Code

Answer: C

Code response: The diameter of the piping constituting the drip shall not be less than that of the line it serves.

Q 15-25 Code requires that drip pipes be provided at certain points to act as storage for condensate. Such drip pipes must be readily accessible for draining and be of _____ inch minimum length.

(A) 10   (C) 14
(B) 12   (D) 18

Standard Gas Code

Answer: B

Code response: The minimum length of the drip piping shall not be less than 12 inches.

Q 15-26 In a building where undiluted liquefied petroleum gas is to be used, the Code prohibits the installation of copper tubing:

(A) in a casing   (C) in hollow partitions
(B) above the floor   (D) inside walls or partitions

Standard Gas Code

Answer: D

Code response: Where undiluted liquefied petroleum gas is to be used, tubing shall not be run inside walls or partitions.

Q 15-27 The <u>one</u> of the following type fittings for an undiluted liquefied petroleum gas piping system which is <u>not</u> Code-approved is:

(A) cast-iron fittings   (C) welded fittings
(B) flanged fittings   (D) brass fittings

Standard Gas Code

Answer: A

Code response: Cast-iron fittings shall be prohibited in an undiluted liquefied petroleum gas piping system.

Q 15-28 According to Code, all pipe or tubing used in LPG gas piping systems must be suitable for a minimum working pressure of _____ pounds per square inch.

(A) 100   (C) 150
(B) 125   (D) 200

Standard Gas Code

Answer: B

Code response: All pipe or tubing (used in an LPG gas piping system) shall be suitable for a working pressure of not less than 125 pounds per square inch.

15-29 The Btu content and specific gravity of undiluted liquefied petroleum gases are such that the provisions covering installation of natural gas piping, according to Code,

(A) are applicable

(B) are not applicable

(C) are very similar

(D) shall be considered the same

Standard Gas Code

Answer: B

Code response: The Btu content and specific gravity of undiluted LPG gases are such that the provisions of Section 305.1, "Other Type Gases," are not applicable.

15-30 Given: The job engineer specifies wrought-steel pipe for a commercial building LPG gas piping system. Due to certain conditions, the gas installer decides to use type-L seamless copper tubing. According to Code, the corresponding size copper tubing to use in lieu of 1-inch specified iron pipe size is ____ inch(es).

(A) 3/4

(B) 7/8

(C) 1-1/8

(D) 1-3/8

Standard Gas Code

Answer: C

Code response: Refer to Table 19 to check corresponding sizes suitable for LPG gases: For iron pipe (nominal size) of 1 inch, the corresponding copper tubing is O.D. 1-1/8 inches.

15-31 The one of the following supporting methods for gas piping <u>not</u> acceptable by Code is:

(A) pipe hooks

(B) other pipes

(C) pipe straps

(D) metal bands

Standard Gas Code

Answer: B

Code response: Gas piping in buildings shall not be supported by other piping.

15-32 The gas installer finds he cannot avoid installing a gas line through a building's heat duct system. According to Code,

(A) there is no objection, if properly supported

(B) such pipe must be accessible for replacement or repairs

(C) the piping material must be resistant to heat

(D) the pipe may be installed in a casing of non-combustible rigid gas-tight material

Standard Gas Code

Answer: D

Code response: Where the Administrative Authority finds that it is impossible to avoid installing gas piping through or in the building duct system for heat, the gas piping may be installed in a casing of noncombustible rigid gas-tight material.

15-33 Code states that immediately after installation, each outlet must be securely closed, gas-tight, and must be left closed until:

(A) all testing has been completed

(B) gas lines have been purged

(C) an appliance is connected to the outlet

(D) accepted by the gas inspector

Standard Gas Code

Answer: C

Code response: Each outlet shall be securely closed gas-tight with a threaded iron plug or cap immediately after installation and shall be left closed until an appliance is connected thereto.

15-34 According to Code, after a gas piping system has been tested, inspected and accepted by local gas inspector, before turning gas into any piping:

(A) the approval of gas system must be recorded with the building and zoning department

(B) a final check by an accurate and sensitive pressure indicating device must be made

(C) all gas outlets must be closed

(D) all gas outlets must be opened

Standard Gas Code

Answer: C

Code response: Before turning gas into any piping, all openings from which gas can escape shall be closed.

15-35 When purging a gas line, Code states that the purged contents shall:

(A) discharge into a dry well

(B) only discharge into the open atmosphere

(C) discharge into the nearest sanitary vent pipe

(D) not discharge into the combustion chamber of an appliance

Standard Gas Code

Answer: D

Code response: Under no circumstances shall a gas line be purged into the combustion chamber of an appliance.

15-36 Changes in direction of gas pipe may be made by the use of bends. According to Code, the pipe can not be bent through an arc of more than _____ degrees.

(A) 45

(B) 60

(C) 90

(D) 120

Standard Gas Code

Answer: C

Code response: Changes in direction of gas pipe may be made by the use of bends. Pipe shall not be bent through an arc of more than 90 degrees.

15-37 When necessary to install piping through concrete or masonry floors or walls above grade, which method of installation is permitted by Code?

(A) Piping may be encased in a 1 to 3 mixture of cement and sand

(B) Opening must be filled to prevent the entrance of corrosive materials

(C) Pipe must be painted with noncorrosive material

(D) Opening must be sealed tight with a flexible material

Standard Gas Code

Answer: A

Code response: When necessary to install piping through concrete or masonry floors or walls above grade, piping may be encased in a 1 to 3 mixture of cement and sand.

15-38 The Code permits the installation of gas piping in a concrete slab when it is not practical to do otherwise. Which one of the following installation methods is prohibited by Code?

(A) Piping may be installed in a channel in the floor

(B) Piping may be installed in concrete floors containing quick-set additives

(C) Piping may be installed in concrete floors constructed with Portland cement

(D) Piping may be installed in a casing of iron pipe with tightly sealed ends

Standard Gas Code

Answer: B

Code response: Gas piping shall not be embedded in concrete slabs containing quick-set additives.

15-39 Compression-type mechanical joints used in gas piping, according to Code, must be designed and installed to effectively sustain the longitudinal pull-out forces caused by:

(A) external loading

(B) gravitational pull

(C) temperatures exceeding 100 degrees

(D) temperatures less than 60 degrees

Standard Gas Code

Answer: A

Code response: Compression-type mechanical joints shall be designed and installed to effectively sustain the longitudinal pull-out forces caused by contraction of the piping or by external loading.

15-40 If you needed to install a restaurant range with 6 top burners and 2 ovens, that has a demand of 150,000 Btu/hr, using a 90-foot run of schedule 40 pipe from point of delivery to range connection, and that is based on a pressure drop of 0.5 inches water column and 0.6 specific gravity gas, the Code-required nominal pipe size would be _____ inch(es).

(A) 1

(B) 1¼

(C) 1½

(D) 2

Standard Gas Code

Answer: A

Code response: Refer to Table 2, "Size of Gas Piping." Based on a pressure drop of 0.5 inches water column, 0.6 specific gravity gas and schedule 40 pipe, length of run is 90 feet and the nominal pipe size of 1 inch will supply 182 cubic feet of gas or approximately 182,000 Btu per hour. Answer "A" is correct.

15-41 The Code states that for conditions such as longer runs or greater gas demands than pipe sizes specified by Code, larger pipe sizes established by the _____ shall be used.

(A) job engineer

(B) consumer

(C) gas company

(D) gas installer

Standard Gas Company

Answer: C

Code response: Where pipe sizes larger than those set forth in the Code are specified by pipe sizing requirements established by the gas company, these larger sizes shall be used.

15-42 The Code states that consumers' gas piping manufactured of wrought-iron or steel pipe must comply with approved standards established by the:

(A) ASTM

(B) IBRM

(C) AGA

(D) ANSI

Standard Gas Code

Answer: D

Code response: Consumers' gas piping shall be wrought-iron or steel pipe complying with the ANSI "Standard for Wrought-Steel and Wrought-Iron Pipe."

15-43 When requested by the Administrative Authority for a particular installation, Code does state that tin-lined copper tubing complying with ANSI/ASTM standard specification:

(A) shall not be used

(B) shall be used

(C) may not be used in temperatures below 32 degrees F.

(D) may be used underground only

Standard Gas Code

Answer: B

Code response: When requested by the Administrative Authority or the gas company, tin-lined copper tubing complying with ANSI/ASTM specifications shall be used.

15-44 Compression-type tubing fittings installed on consumers' gas piping, according to Code, may be used:

(A) for outside underground connections

(B) for outside above ground connections

(C) in hollow partitions

(D) in solid partitions

Standard Gas Code

Answer: A

Code response: Compression-type tubing fittings shall not be used inside or under buildings, but may be used for other underground connections.

15-45 The connection between the metallic piping and the plastic piping, according to Code, must be made with a plastic-to-steel transition fitting:

(A) if first approved by Administrative Authority

(B) provided plastic pipe is polyvinyl chloride, type II

(C) as recommended by the plastic piping manufacturer

(D) as recommended by the steel pipe manufacturer

Standard Gas Code

Answer: C

Code response: The connection between the metallic piping and the plastic piping shall be a mechanical joint type, or plastic-to-steel transition fitting as recommended by the plastic piping manufacturer.

15-46 An inspector arrives to inspect a gas installation in a commercial building. The use of an extension ladder is necessary. He is unable to complete the inspection because:

(A) he forgot to bring his ladder

(B) the general contractor is using the ladder on another job

(C) the architect failed to specify who would supply job ladders

(D) the gas installer has loaded all tools and equipment on his truck and left the job site

Standard Gas Code

Answer: D

Code response: For inspections, all tools, apparatus, labor and assistance necessary for the tests shall be furnished by the installer.

15-47 According to Code, the inspection of all new gas piping, which includes a pressure test, is a:

(A) rough piping inspection

(B) final piping inspection

(C) final inspection

(D) partial inspection

Standard Gas Code

Answer: B

Code response: Final piping inspection shall be made after all piping authorized by the permit has been installed, after piping to be concealed has been concealed, and before any fixtures or gas appliances have been attached. This inspection shall include a pressure test.

15-48 Code states that responsibility for meeting Code requirements in consumers' gas piping installation rests with the:

(A) consumer

(B) inspector

(C) installer

(D) architect

Standard Gas Code

Answer: C

Code response: The responsibility for observing Code requirements shall rest with the installing agency.

15-49 When the only practical manner of installing gas piping is to have it pass through the outer foundation wall of a building from below grade outside to above grade inside, Code requires that the pipe be sleeved and sealed to prevent:

(A) entry of water

(B) entry of bugs or rodents

(C) possibility of gas seeping into the building

(D) corrosion of pipe

Standard Gas Code

Answer: A

Code response: Gas piping, when installed to pass through the outer foundation wall of a building from below grade outside to above grade inside, shall be encased in a sleeve and sealed at the outside of the foundation wall to prevent entry of water.

15-50 Of the following, the most important prerequisite of the Gas Code is that when installing gas piping it must:

(A) be graded to properly drain condensate to a container

(B) be properly supported so it cannot be accidentally moved

(C) not be in contact with the ground under a building slab

(D) conform with good workmanship

Standard Gas Code

Answer: C

Code response: Buried house piping shall not be installed in such a way as to be in contact with the ground or fill under a building or floor slab.

Chapter 16

Gas Appliances, Vents and Flues

As a plumber or gas installer, you'll install many gas-fired water heaters, ranges, dryers and room heaters. Many plumbers learn this work by simply watching others — without ever understanding what the code says or why the work is done like it is. But to pass the exam, you have to know the details, what the code says about installation and venting of gas appliances.

Appliances

This section covers the basic requirements for installation of specific types of appliances. The code sets the minimum clearance from combustible construction, identifies prohibited installations, regulates accessibility, and mandates the use of manual and automatic shut-off valves, for exam-ple. Responsibility for observing these requirements always rests with the installer.

Appliance Venting

The code establishes two classes of gas appliances, those requiring vents and those not requiring vents. The code states: "Venting systems shall be engineered and constructed so as to develop a positive flow adequate to remove flue gases safely and to the outside atmosphere."

Materials used for vent or flue piping construction, the type of vents required for different gas appliances and the installation methods are all specified in the code. Again, the responsibility for observing these requirements rests with the installer.

16-1 According to Code, in a natural gas system where the required flow rate is 3.0 cubic feet per hour and the pressure at the fixed orifice of an appliance is 10.0 inches water column, the orifice acceptable for use is the one having a gas flow of:

(A) 79

(B) 72

(C) 70

(D) 69

Standard Gas Code

Answer: B

Code response: Refer to Table B-1, Appendix B. "Flow of Gas Through Fixed Orifices."

16-2 Gas appliances installed in the attic of a building, according to Code, must:

(A) be protected with an automatic fire protection system

(B) be provided with a continuous floored passageway at least 2 feet wide

(C) be provided with a floor of a fireproof material

(D) be vented through the roof with a single wall flue pipe

Standard Gas Code

Answer: B

Code response: For attic installation, such attics shall have flooring under and around such gas appliances and shall be provided with a continuous floored passageway at least 2 feet wide.

16-3 When a gas water heater is installed in a separate room, ventilation must be provided through permanent openings. According to Code, one vent opening should be a 12-inch minimum above the floor, the second opening should be a 12-inch minimum below the ceiling. Assuming the gas water heater has an input rating of 250,000 Btu per hour, the total free area of the permanent openings should be _____ square inches.

(A) 100

(B) 250

(C) 350

(D) 500

Standard Gas Code

Answer: D

Code response: The two permanent openings shall each have a free area of not less than 1 square inch per 1,000 Btu per hour of total input rating of all appliances in the confined space.

Solution: $\dfrac{250,000 \text{ Btu per hour}}{1,000 \text{ per sq. in.}}$ = 250 sq. in. per opening x 2 openings = 500 sq. in. for both openings.

Q **16-4** According to Code, the heat required for a restaurant range having **6** top burners and **4** ovens would be approximately _____ Btu/hour.

(A) 200,000

(B) 225,000

(C) 240,000

(D) 250,000

Standard Gas Code

Answer: C

Code response: The approximate maximum gas demand for a restaurant range having 6 top burners and 4 ovens is 240,000 Btu/hour. Refer to Table 1, "Approximate Maximum Demand of Typical Gas Appliances in Btu per Hour."

Q **16-5** You are installing gas piping for ironing equipment that requires mobility during operation. According to Code, a gas hose may be used, provided the length, "X" in Figure 16-1, does not exceed _____ feet.

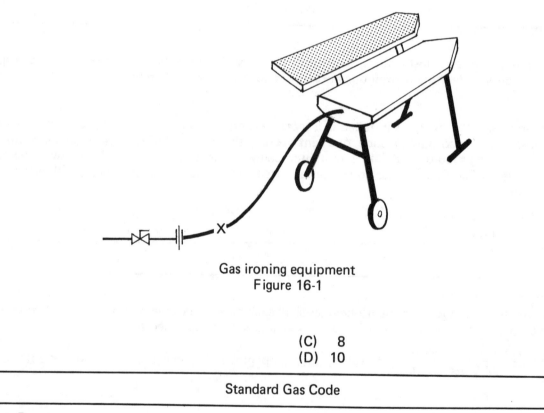

Gas ironing equipment
Figure 16-1

(A) 4

(B) 6

(C) 8

(D) 10

Standard Gas Code

Answer: B

Code response: Where the gas hose is used for equipment requiring mobility during operation, it shall be of the minimum practical length, but not to exceed 6 feet.

16-6 The Code requirement for a water heater with an electric ignition system is that it must ignite the pilot within _____ seconds after the gas supply is turned on.

(A) 5

(B) 10

(C) 15

(D) 20

Standard Gas Code

Answer: C

Code response: The gas shall shut off automatically if the pilot is not ignited within 15 seconds by an electrical ignition system.

16-7 According to Code, gas-fired water heaters with electric ignition systems are permitted to ignite:

(A) only a pilot

(B) a natural-type draft burner only

(C) an LPG gas burner only

(D) a natural gas burner only

Standard Gas Code

Answer: A

Code response: Electric ignition systems shall ignite only a pilot.

16-8 The approximate maximum Btu per hour input for a commercial gas range having 4 top burners and 2 ovens, according to Code, is _____ Btu per hour.

(A) 150,000

(B) 185,000

(C) 220,000

(D) 240,000

Standard Gas Code

Answer: A

Code response: The approximate maximum gas demand for a restaurant range having 4 top burners and 2 ovens is 150,000 Btu/hour. Refer to Table 1, "Approximate Maximum Demand of Typical Gas Appliances in Btu per Hour."

16-9 According to Code, a central gas-fired heating boiler utilizing complete shut-off type safety devices must be provided with a _____ ahead of all controls.

(A) manual shut-off valve

(B) drip valve

(C) pressure gauge

(D) relief valve

Standard Gas Code

Answer: A

Code response: Where a complete shut-off type safety device is utilized, a manual shut-off valve shall be provided ahead of all controls.

16-10 Where gas appliances are installed in a confined space, provisions must be made to supply this space with air for combustion, ventilation and dilution of flue gases. According to Code, <u>each</u> of the permanent openings, marked "A" and "B," as illustrated in Figure 16-2, must have a minimum free area of ____ square inches.

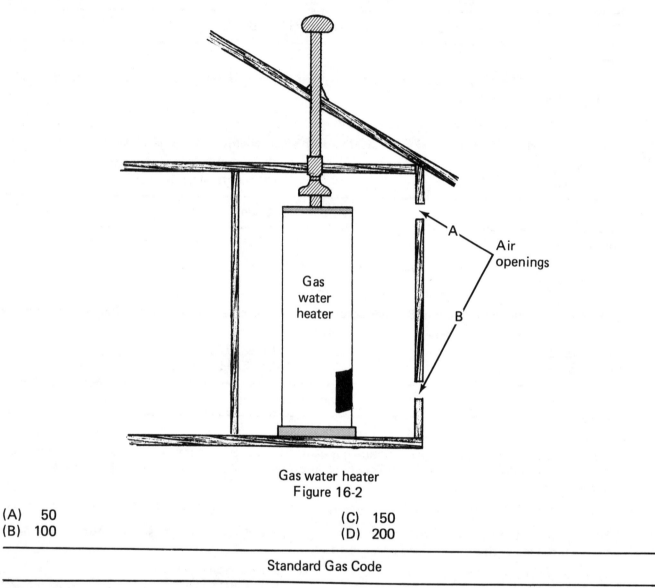

Gas water heater
Figure 16-2

(A) 50

(B) 100

(C) 150

(D) 200

Standard Gas Code

Answer: B

Code response: Where gas appliances are installed in a confined space, the two permanent openings shall **each** have a free area of not less than 1 square inch per 1,000 Btu per hour of total input rating, but not less than 100 square inches.

16-11 According to Code, the control side of a suspended type gas-fired unit heater must be spaced at least ____ inches from a wall.

(A) 6

(B) 10

(C) 16

(D) 18

Standard Gas Code

Answer: D

Code response: The control side of a unit heater shall be spaced not less than 18 inches from any wall.

16-12 Natural gas releases approximately 1,000 Btu per cubic foot. According to the Gas Code, a gas water heater having a maximum demand of 230,000 Btu/hour will consume approximately ____ cubic feet of gas each ½ hour.

(A) 115

(B) 150

(C) 175

(D) 230

Standard Gas Code

Answer: A

Solution:

Natural gas demand = $\dfrac{230,000 \text{ Btu/hour}}{1,000 \text{ Btu/cu. ft.}}$ = 230 cu. ft./hr. or ½ of 230 cu. ft. = 115 cu. ft. for each ½ hour

16-13 When installing suspended type gas-fired unit heaters in garages where four motor vehicles will use the facilities, Code requires that the unit heaters be installed a minimum of ____ feet above the floor.

(A) 7

(B) 8

(C) 9

(D) 10

Standard Gas Code

Answer: B

Code response: Unit heaters installed in garages for more than 3 motor vehicles shall be at least 8 feet above the floor.

16-14 According to Code, unvented room heaters may be installed if they are:

(A) installed in accordance with their listing

(B) equipped with an oxygen-depletion sensitive safety shut-off system

(C) installed according to the manufacturer's instructions

(D) not attached to the building

Standard Gas Code

Answer: B

Code response: All unvented room heaters shall be equipped with an oxygen-depletion sensitive safety shut-off system.

16-15 Gas room heaters installed in corridors of homes for the aged:

(A)   must be of the vented type

(B)   may be of the unvented type

(C)   must be a minimum of 8 inches from walls of combustible construction

(D)   shall not have projecting flue box

Standard Gas Code

Answer: A

Code response: Gas room heaters installed at any location in homes for the aged shall be of the vented type.

16-16 When installing listed, unvented gas-fired room heaters, the Code states that the normal input rating cannot exceed _____ Btu/hour.

(A)   20,000

(B)   30,000

(C)   40,000

(D)   50,000

Standard Gas Code

Answer: C

Code response: Listed, unvented gas-fired room heaters shall not have a normal input rating in excess of 40,000 Btu per hour.

16-17 Code prohibits installation of a listed unvented gas-fired room heater in:

(A)   a residential living room

(B)   an enclosed porch

(C)   a hallway between two bedroom doors

(D)   a bedroom

Standard Gas Code

Answer: D

Code response: Listed unvented gas-fired room heaters shall not be installed in sleeping quarters.

16-18 According to Code, when installing an underfired, gas automatic water heater, the heater's jacket must be at least _____ inch(es) from any combustible material.

(A)   1

(B)   2

(C)   4

(D)   6

Standard Gas Code

Answer: B

Code response: Minimum clearance from the nearest part of the gas-fired automatic water heater jacket to the combustible material shall be 2 inches.

Q 16-19 According to Code, a ventilating hood must be provided above an open-top broiler unit. The minimum clearance between the cooking top and combustible material above the hood is _____ inches.

(A) 18

(B) 20

(C) 22

(D) 24

Standard Gas Code

Answer: D

Code response: A ventilating hood shall be provided above an open-top broiler unit. A minimum clearance of 24 inches shall be maintained between the cooking top and combustible material above the hood.

Q 16-20 You are replacing an old range with a modern gas range. According to Code, the gas:

(A) pipe connection should be of rigid material

(B) should be turned off

(C) may be left on temporarily for the exchange

(D) line should be purged before connecting the new range

Standard Gas Code

Answer: B

Code response: All gas appliance installation shall be done with the gas turned off to eliminate hazards from leakage of gas.

Q 16-21 A water heater manufactured to use natural gas is shipped to a job where LPG is the only gas available. According to Code, the gas heater:

(A) may be installed if it complies with ANSI Standard Requirements

(B) shall have a type "A" gas vent installed

(C) orifice must be changed for the different type gas

(D) shall not be installed

Standard Gas Code

Answer: D

Code response: No attempt shall be made to convert the appliance from the gas specified on the rating plate for use with a different gas.

Q 16-22 Equipment that requires mobility during operation may be connected to a gas hose. However, the Code specifies that a gas hose cannot be used:

(A) if the temperature of the room is more than 125 degrees F

(B) if the temperature of the room is less than 75 degrees F

(C) unless first approved by Administrative Authority

(D) if the supply piping is larger than the hose specified

Standard Gas Code

Answer: A

Code response: Gas hose shall not be used where it is likely to be subject to excessive temperatures, above 125 degrees F.

Q 16-23 When installing a gas range, the Code mandates that an accessible gas shut-off valve be installed upstream from the union or range connector. Distance "X" in Figure 16-3 cannot exceed _____ feet.

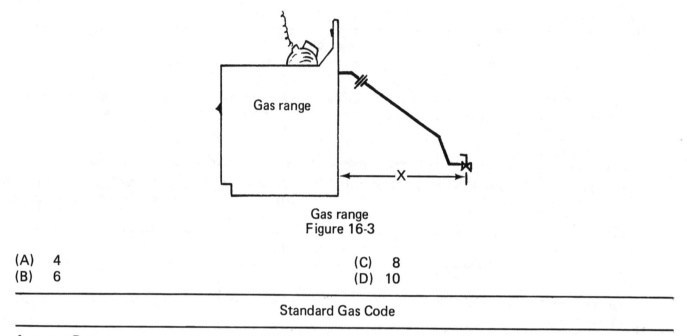

Gas range
Figure 16-3

(A) 4	(C) 8
(B) 6	(D) 10

Standard Gas Code

Answer: B

Code response: All gas appliances shall have accessible gas shut-off valves located no further than 6 feet from the appliance.

Q 16-24 Portable outdoor gas-fired appliances may be connected with gas hose connectors listed for that purpose. Length of hose "X" in Figure 16-4 must be no more than _____ feet, if it is to meet Code specifications.

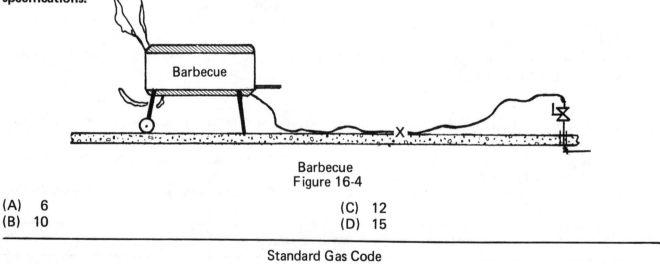

Barbecue
Figure 16-4

(A) 6	(C) 12
(B) 10	(D) 15

Standard Gas Code

Answer: D

Code response: The length of assembled connectors for portable outdoor gas-fired appliances shall be limited to 15 feet.

16-25 Sometimes a situation exists where a gas water heater must be installed in a bedroom. The Code allows for such an exceptional location, provided:

(A) an oxygen depletion safety shut-off system is stipulated

(B) the input rating of water heater is less than 25,000 Btu/hour

(C) the water heater has a sealed combustion system

(D) air for combustion and ventilation is adequate

Standard Gas Code

Answer: C

Code response: Water heaters, with the exception of those having sealed combustion systems, shall not be installed in bedrooms.

16-26 The Code requires that gas appliances be installed in a location in which the ventilation facilities permit satisfactory combustion of gas and proper venting under normal conditions. A gas water heater and furnace are to be installed in an unconfined space. All of the following building construction materials provide acceptable ventilation conditions <u>except</u>:

(A) wood frame

(B) brick

(C) stucco block

(D) stone

Standard Gas Code

Answer: C

Code response: Where appliances are installed in unconfined spaces in buildings of conventional frame, brick, or stone construction, infiltration is normally adequate to provide air for combustion and draft hood dilution.

16-27 Gas appliances using electrical controls, according to Code, shall have the controls:

(A) connected into a permanently live electrical circuit

(B) connected to a light switch

(C) approved by the electrical inspector before installation

(D) connected directly to the house panel

Standard Gas Code

Answer: A

Code response: All gas appliances using electrical controls shall have the controls connected into a permanently live electrical circuit, i.e., one that is not controlled by a light switch.

16-28 When installing a gas appliance on the roof of a building that does not have parapet walls at least 3 feet high, the Code requires that the appliance be set at least _____ feet from the edge of the roof.

(A) 2
(B) 4

(C) 6
(D) 8

Standard Gas Code

Answer: C

Code response: Appliances installed on the roof of a building that does not have rails, guards or parapet walls at least 3 feet in height shall have at least 6 feet clearance from the roof edge.

16-29 Code mandates that a means of interrupting the electrical supply to an air conditioning appliance be installed within sight of, but not over _____ feet from the air conditioner.

(A) 30
(B) 50

(C) 75
(D) 100

Standard Gas Code

Answer: B

Code response: Means for interrupting the electrical supply to an air conditioning appliance shall be provided within sight of and not over 50 feet from the air conditioner.

16-30 According to Code, gas-fired floor furnaces must be protected, where necessary:

(A) from vehicular traffic
(B) against severe wind conditions

(C) to prevent gravity circulation
(D) against anti-siphoning conditions

Standard Gas Code

Answer: B

Code response: Gas floor furnaces shall be protected, where necessary, against severe wind conditions.

16-31 According to Code, natural draft gas vents, installed through and terminating adjacent to an outside wall:

(A) shall terminate no less than 12 inches from the wall
(B) shall terminate no less than 24 inches from the wall

(C) must be provided with a rain collar
(D) are prohibited

Standard Gas Code

Answer: D

Code response: Natural draft vents extending through and terminating adjacent to outside walls are prohibited.

Q 16-32 Given: A 500,000 Btu/hour input single gas appliance is installed with a double-wall metal flue pipe. Vertical distance from draft hood to vent cap is 20 feet. Lateral distance is zero feet. The Code requires that the flue pipe size used for the gas appliance be _____ inches.

(A) 4   (C) 6
(B) 5   (D) 8

Standard Gas Code

Answer: D

Code response: A 500,000 Btu/hour input single gas appliance installed according to the above question requirements shall have a vent diameter of 8 inches. Refer to Table 12, "Capacity of Type B Double-Wall Vents with Type B Double-Wall Connectors Serving a Single Appliance."

Q 16-33 According to Code, exit terminals of mechanical draft systems must be located a minimum of _____ feet above grade when adjacent to public walkways.

(A) 6   (C) 8
(B) 7   (D) 9

Standard Gas Code

Answer: B

Code response: Exit terminals of mechanical draft systems shall be not less than 7 feet above grade when located adjacent to public walkways.

Q 16-34 Code states that single-wall metal vent pipes may pass through combustible walls or partitions, provided they are protected at the point of passage by:

(A) ventilated metal thimbles   (C) metal sleeves
(B) metal stud guards   (D) asbestos millboard

Standard Gas Code

Answer: A

Code response: Single-wall metal vent connectors shall not pass through any combustible walls unless they are guarded at the point of passage by ventilated metal thimbles.

Q 16-35 A gas vent pipe which passes through the roof of a building must be provided with a vent cap. According to Code, it must not obstruct or reduce the effective cross-sectional area of the vent outlet and can be used on all but a:

(A) type-L vent   (C) type-BW vent
(B) type-C vent   (D) type-B vent

Standard Gas Code

Answer: B

Code response: A vent cap which does not obstruct or reduce the effective cross-sectional area of the vent outlet shall be used on all type-B, BW and type-L vents.

Q 16-36 The vertical height of vent pipe "B" in Figure 16-5 is 20 feet, and vents one gas water heater. The vent pipe is type-L, double-wall. According to Code, the horizontal vent run "A" must not exceed _____ feet.

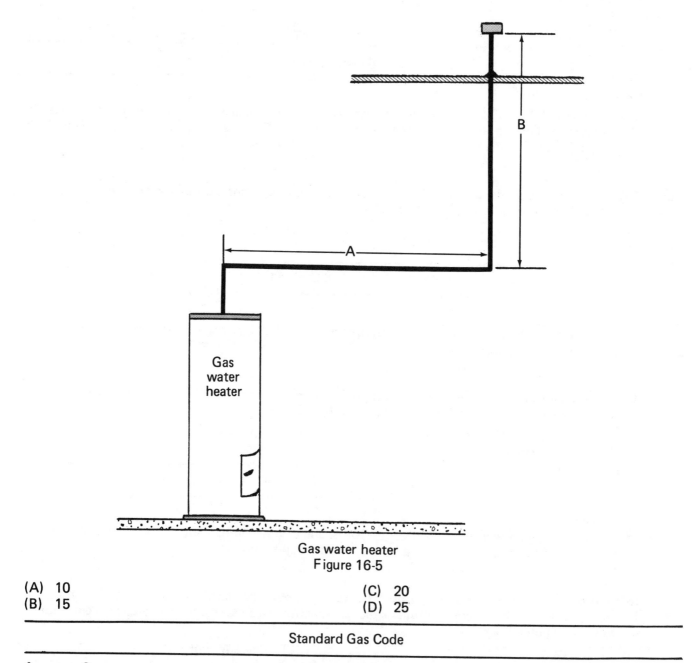

Gas water heater
Figure 16-5

(A) 10
(B) 15

(C) 20
(D) 25

Standard Gas Code

Answer: C

Code response: The maximum horizontal length of a type-L double-wall vent connector, venting one appliance, shall not exceed 100 percent of the height of the vertical vent.

Q 16-37 The vent pipe shown in Figure 16-6 is of single-wall type construction and vents one gas-fired water heater. According to Code, the horizontal section "A" cannot exceed _____ percent of the vertical section "B."

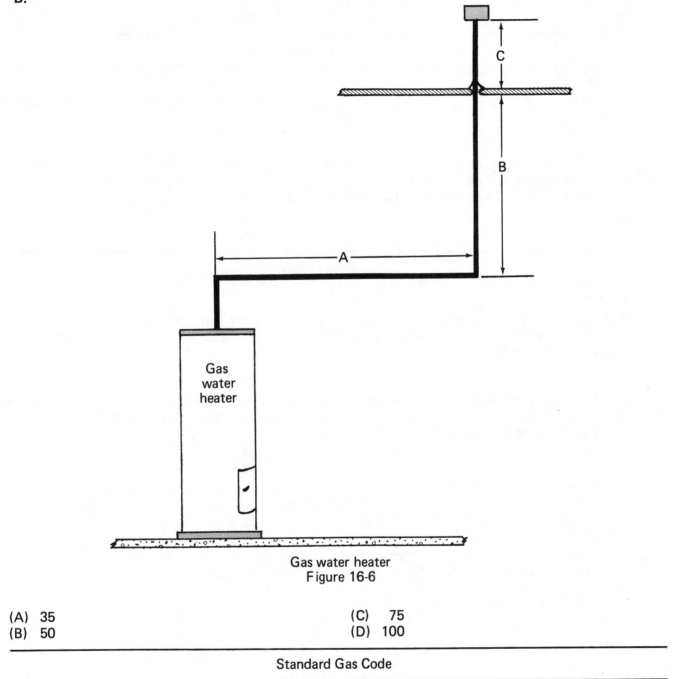

Gas water heater
Figure 16-6

(A) 35   (C) 75
(B) 50   (D) 100

Standard Gas Code

Answer: C

Code response: The length of a horizontal single-wall vent connector venting one appliance shall not exceed 75 percent of the vertical height of the gas vent.

Q. 16-38 Refer again to Figure 16-6. The Code requires that distance "C" vent pipe must extend a minimum of ____ inches above the highest point where it passes through the roof.

(A) 18   (C) 24
(B) 20   (D) 30

Standard Gas Code

Answer: C

Code response: Gas vents shall extend at least 2 feet above the highest point where they pass through the roof.

Q. 16-39 Given: A gas appliance is continually in operation in a room where the average temperature never exceeds 78 degrees F. Code requires that the vent pipe be installed in such a manner that the total temperature surrounding the combustible construction should not exceed ____ degrees F.

(A) 90   (C) 154
(B) 110   (D) 168

Standard Gas Code

Answer: D

Code response: Vent connectors shall be located in such a manner that continued operation of the appliance will not raise the temperature of the surrounding combustible construction to more than 90 degrees **above normal room temperature.**

Q. 16-40 You are installing a 6-inch horizontal single-wall vent pipe. The distance from the appliance to the vertical portion of the vent connection is 15 feet. The Code requires a minimum pitch for the horizontal vent pipe of no less than ____ inches per foot.

(A) .75   (C) .25
(B) .5   (D) .125

Standard Gas Code

Answer: C

Code response: Horizontal vent connectors shall maintain a pitch or rise from the appliance to the vent. A rise as great as possible, at least ¼-inch to the foot (horizontal length), shall be maintained.

16-41 Certain single-wall vent piping (for gas appliances having draft hoods, equipped with listed conversion burners, not installed in attics) must be constructed of materials resistant to corrosion and heat. According to Code the sheet steel used in manufacturing such vent piping must be at least _____ gauge.

(A) 20

(B) 22

(C) 24

(D) 28

Standard Gas Code

Answer: D

Code response: Single-wall vent connectors used for gas appliances having draft hoods and equipped with listed conversion burners which are not installed in attics shall be constructed of materials having a resistance to corrosion and heat not less than that of No. 28 manufacturer's gauge galvanized sheet steel.

16-42 A single-wall metal flue pipe installed to serve a gas-fired water heater is required by Code to have a minimum clearance of _____ inches from any combustible type material.

(A) 2

(B) 4

(C) 6

(D) 8

Standard Gas Code

Answer: C

Code response: Water heater single-wall vent connectors shall have a minimum clearance from combustible construction of 6 inches. Refer to Table 9, "Draft Hood and Vent Connector Clearance For Listed Appliances, unless Otherwise Marked."

16-43 A gas-fired water heater may be installed on the floor of a residential garage. According to Code, if the adjacent ground level is not below or level with the garage floor door, the appliance must be installed at least _____ inches above the floor.

(A) 12

(B) 18

(C) 24

(D) 30

Standard Gas Code

Answer: B

Code response: Gas appliances may be installed on the floor of a residential garage provided a door of the garage opens to an adjacent ground or driveway level that is at or below the level of the garage floor. Where this condition does not exist, appliances shall be installed not less than 18 inches above the floor.

16-44 You are installing vent piping for a single gas-burning piece of equipment with a draft hood diameter of 12 inches. According to Code, the vent pipe used must have a minimum diameter of _____ inches.

(A) 8

(B) 10

(C) 12

(D) 14

Standard Gas Code

Answer: B

Code response: Vents for draft hoods 12 inches in diameter or less should not be reduced more than one size. (12 inches to 10 inches is a one-size reduction.)

16-45 Given: The calculated equivalent inside area of a gas-fired appliance requiring a vent 18 feet in height is 607 square inches. According to Code, the minimum nominal liner size, then, is:

(A) 20" x 20"

(B) 24" x 24"

(C) 28" x 28"

(D) 30" x 30"

Standard Gas Code

Answer: D

Code response: Refer to Table 17, "Masonry Chimney Liner Dimensions with Circular Equivalents," and find the appropriate line.

16-46 When installing an automatically-operated gas-fired water heater vented through a natural draft ventilating hood, Code prohibits the installation of a ____ in the ventilating system.

(A) damper

(B) thermostat

(C) flue exhauster

(D) draft regulator

Standard Gas Code

Answer: A

Code response: When automatically-operated gas appliances such as water heaters are vented through natural draft ventilating hoods, dampers shall not be installed in the ventilating system.

16-47 Sometimes it becomes necessary to connect two gas appliances to one vent. The Code specifies the insurance of better draft conditions by having the appliance joined to the vent:

(A) at the same level

(B) at different levels

(C) through a common connector only

(D) through an asbestos cement Type "B" connector only

Standard Gas Code

Answer: B

Code response: In order to promote better draft where more than one gas appliance vent connector is connected to a vent, the connection should be made at different levels.

16-48 Code mandates that a vent pipe for a gas-fired water heater with a 4-inch diameter draft hood outlet connect with a vent pipe whose diameter is at least ____ inches.

(A) 3

(B) 3½

(C) 4

(D) 4½

Standard Gas Code

Answer: C

Code response: The vent connector shall not be smaller than the size of the outlet of the draft hood supplied by the manufacturer of a gas-designed appliance.

Q 16-49 Of the following single-wall appliance vent locations, all are prohibited by Code except for a vent installed:

(A) horizontally in a plastered partition

(B) vertically through the attic to the outer air

(C) vertically through the floor to the roof and to the outer air

(D) through the exterior wall to the outer air

Standard Gas Code

Answer: D

Code response: Single-wall vents shall be used only for runs directly from the space in which the appliance is located through the roof or exterior wall to the outer air. Such vents shall not be installed in any attic or concealed space nor through any floor.

Q 16-50 According to Code, every appliance listed for vented use must have a:

(A) ventilated metal thimble

(B) draft hood

(C) elliptical type vent

(D) adjustable (modulating) damper

Standard Gas Code

Answer: B

Code response: Every appliance listed for vented use shall have a draft hood.

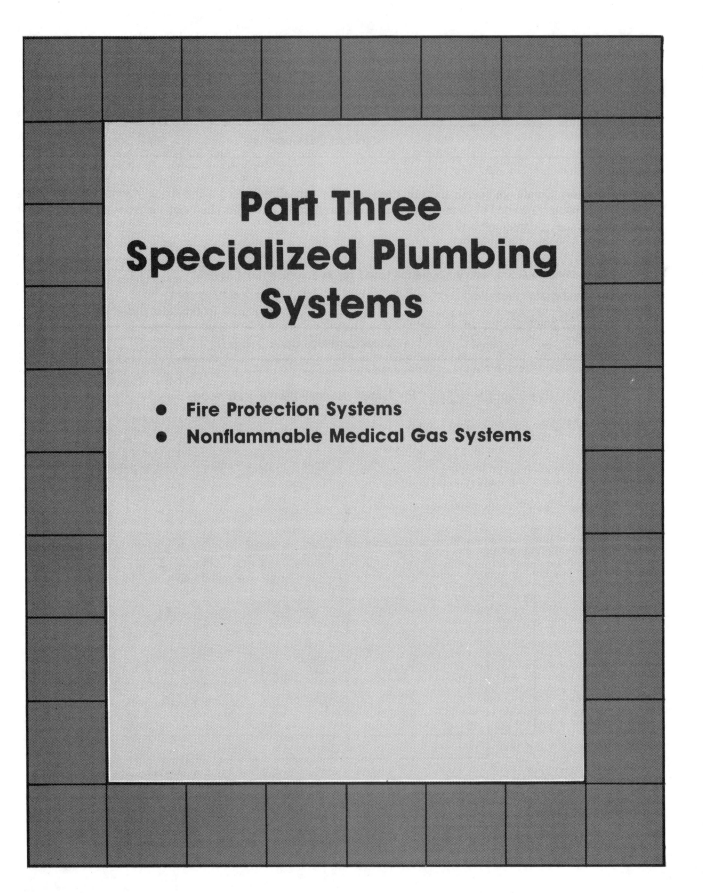

Part Three
Specialized Plumbing Systems

- Fire Protection Systems
- Nonflammable Medical Gas Systems

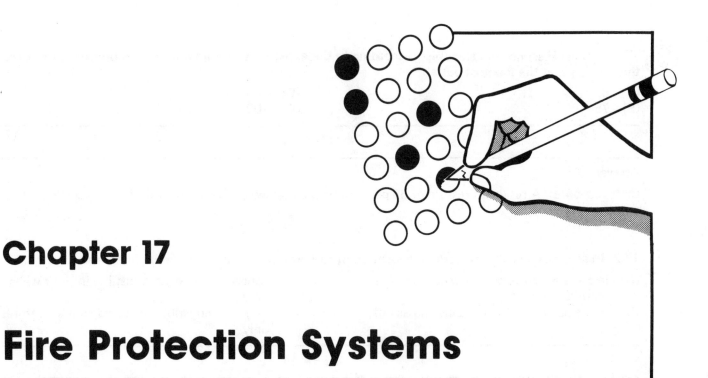

Chapter 17

Fire Protection Systems

Fire protection systems are required in many buildings. Installing the system is usually considered part of a plumber's job. You can expect questions on fire protection systems (standpipes and fire hoses) on your examination. This section of the test will cover standards for materials used, installation practice, and testing methods.

In some states, anyone who installs automatic fire sprinklers must be certified by the state fire marshall. If this certification is required, the fire marshall will administer the sprinkler installer's exam and the plumber's exam will not cover automatic fire sprinklers.

17-1 A Class II service fire standpipe, according to Code, requires a minimum residual pressure of not less than _____ psi at the topmost outlet.

(A) 65

(B) 75

(C) 85

(D) 100

NFPA 14

Answer: A

Code response: A residual pressure of 65 psi shall be maintained at the topmost outlet of each Class II service standpipe.

17-2 In accordance with the Code, the location of fire protection standpipes in buildings must be:

(A) in areas constructed of noncombustible materials only

(B) in specially constructed pipe chases only

(C) concealed in noncombustible interior walls only

(D) in noncombustible fire-rated stair enclosures only

NFPA 14

Answer: D

Code response: Fire protection standpipes shall be located in noncombustible fire-rated stair enclosures.

17-3 According to Code, fire protection standpipe systems for multi-story buildings exceeding 275 feet in height must be:

(A) sized by hydraulic calculations

(B) zoned accordingly

(C) designed with 6" minimum size piping

(D) designed by the local authority

NFPA 14

Answer: B

Code response: Buildings in excess of 275 feet in height shall be zoned accordingly.

17-4 As outlined by Code, fire standpipe risers must be designed so that a stream of water can be brought to bear on all parts of all floors within 30 feet of a nozzle, which in turn is connected to not more than _____ feet of riser-attached hose.

(A) 50

(B) 75

(C) 100

(D) 125

NFPA 14

Answer: C

Code response: All portions of each story of a building shall be within 30 feet of a nozzle when attached to not more than 100 feet of hose.

17-5 Of the following occupancies, the one classified by Code as "light hazard" is:

(A) machine shops
(B) mercantiles

(C) feed mills
(D) museums

NFPA 14

Answer: D

Code response: Light hazard occupancies are defined as buildings where the quantity and/or combustibility of contents is low. Museums are thus classified.

17-6 According to Code, fire standpipes for Class II service shall be provided with _____ hose connections on each floor.

(A) 1¼-inch
(B) 1½-inch

(C) 2-inch
(D) 2½-inch

NFPA 14

Answer: B

Code response: For Class II service, fire standpipes shall be provided with 1½-inch hose connections on each floor.

17-7 The minimum base diameter of an installed fire standpipe in a 16-story office building, according to Code, cannot be smaller than:

(A) 4 inches
(B) 5 inches

(C) 6 inches
(D) 8 inches

NFPA 14

Answer: C

Code response: A fire standpipe in excess of 100 feet in height shall be a minimum of 6 inches in size at its base.

17-8 The jurisdictional authority can, by Code, require that high-rise buildings under construction have:

(A) wet standpipes with at least one fire department hose valve at each floor level
(B) no active fire protection systems during construction

(C) a minimum 600-gallon water storage tank at highest completed floor
(D) a hand-operated hydraulic water pump readily available

NFPA 14

Answer: A

Code response: At least one fire department hose valve shall be provided at each floor level for fire department use for buildings under construction.

17-9 According to Code, a building with nine standpipes, conforming to Class III fire protection service must have an automatic fire pump providing _____ gpm minimum flow.

(A) 2,000   (C) 2,500

(B) 2,350   (D) 2,750

NFPA 14

Answer: C

Code response: In standpipe systems for Class III buildings, all supply piping shall be sized for minimum flow. First standpipe is 500 gpm, each additional standpipe is 250 gpm.

Solution:
1 standpipe x 500 gpm = 500 gpm
8 standpipes x 250 gpm = 2,000 gpm
Total 2,500 gpm for all 9 standpipes

17-10 The Code states that a 750 gpm capacity fire pump for a standpipe system may be provided with _____ 2½-inch hose wall outlet(s) at ground level for fire department use.

(A) 1   (C) 3

(B) 2   (D) 4

NFPA 14

Answer: C

Code response: One 2½-inch hose outlet may be provided in the form of a fire department wall outlet at ground level for each 250 gpm capacity fire pump for a standpipe system.

Solution: 3 x 250 gpm/fire pump = 750 gpm capacity

17-11 According to Code, required fire system standpipes 30 feet high must have a minimum diameter of _____ inches.

(A) 1½   (C) 2½

(B) 2   (D) 3

NFPA 14

Answer: B

Code response: Fire standpipes under 50 feet in height shall be a minimum of 2 inches in diameter.

Note: The minimum size for fire standpipes acceptable by most local codes is 4 inches in diameter.

17-12 According to Code, a Class I service fire standpipe system of 5 standpipes, with a maximum 2,500 gpm flow, requires a minimum residual pressure of _____ psi at its most remote roof manifold outlet.

(A) 65
(B) 75

(C) 100
(D) 125

NFPA 14

Answer: A

Code response: For Class I service, the minimum supply shall be sufficient to maintain a residual pressure of 65 psi at the topmost outlet of the most remote standpipe.

17-13 According to the Code, an unlined fire hose shall be provided with a:

(A) soft-seat check valve
(B) solenoid valve

(C) listed automatic drip connection
(D) NFPA approved fire truck connector

NFPA 14

Answer: C

Code response: Hose valves attached to unlined fire hoses shall be provided with a listed automatic drip connection.

17-14 A building requiring standpipes and fire hose cabinets for emergency use must, according to Code, have fire hoses of at least _____ inches.

(A) 1¼
(B) 1½

(C) 1¾
(D) 2

NFPA 14

Answer: B

Code response: Racks in fire hose cabinets shall have storage facilities for 1½-inch hose for use by occupants in an emergency.

17-15 Given: A building is completely sprinklered. The fire protection is a combined system design. The risers are sized by hydraulic calculations. Under these conditions, Code requires that the contractor:

(A) if requested by local jurisdictional authority, submit complete hydraulic calculations
(B) install gate valves at the base of each standpipe riser

(C) test all portions of the system under actual "fire" simulated conditions
(D) weld all joints 2 inches and larger

NFPA 14

Answer: A

Code response: In buildings with combined systems, designed to be completely sprinklered, with risers sized by hydraulic calculations, the contractor must submit complete calculations when requested by the authority having jurisdiction.

17-16 Of the following statements, which one is applicable to fire standpipe systems, according to Code?

(A) Each floor sprinkler system must operate independently of the fire standpipes

(B) In a high-rise building having three fire standpipes, sprinkler systems at each floor level shall connect only to one approved fire standpipe

(C) At least one water supply must be automatic

(D) A public water system cannot be used to supply water for a fire standpipe system

---

NFPA 14

Answer: C

Code response: Fire standpipe systems shall have at least one water supply that is automatic.

17-17 The classification "Ordinary Hazard" occupancies (Group 3) does not, according to Code, include:

(A) feed mills

(B) piers and wharves

(C) restaurant service areas

(D) wood machining businesses

NFPA 14

Answer: C

Code response: Ordinary Hazard (Group 3) occupancies are defined as buildings where the quantity and/or combustibility of the contents is high. Restaurant service areas are not thus classified.

17-18 According to Code, when a fire protection system is to operate at 200 psi, it must be tested at a minimum pressure of _____ psi.

(A) 160

(B) 180

(C) 200

(D) 250

NFPA 14

Answer: D

Code response: A new fire system including site piping shall be tested hydrostatically at not less than 200 psi, or at 50 psi in excess of the normal pressure when the normal pressure exceeds 150 psi.

17-19 Dry standpipe systems approved for installation by the jurisdictional authority must be hydrostatically tested at 50 psi above normal working pressure. Code mandates that such hydrostatic test be performed during the life of the system at not less than _____ year intervals.

(A) 1

(B) 3

(C) 5

(D) 7

NFPA 14

Answer: C

Code response: Fire standpipe systems which normally remain dry shall be hydrostatically tested at 50 psi above the normal pressure, at intervals not less than 5 years.

17-20 Sometimes, during modifications to an existing fire protection system, it becomes necessary to remove all or a portion of the system from service. When this happens, the Code states that it is your responsibility to:

(A) notify the local plumbing department

(B) notify the local fire department

(C) post a sign to that effect on each fire department connection

(D) notify the local health department before the work begins

NFPA 14

Answer: B

Code response: Notice shall be given to the local fire department when all or any portion of a standpipe system is out of service for any reason.

17-21 The fire standpipe is located in a stairway enclosure of a 10-story office building. The hose connection on each floor, according to Code, should be installed and ready for use in all of the following locations **except**:

(A) the corridor adjacent to the stairway enclosure

(B) the other space adjacent to the stairway enclosure

(C) the stairway enclosure close to the standpipe

(D) on the wall outside of the pipe shaft housing the fire standpipe

NFPA 14

Answer: C

Code response: Hose connections for Class II service should be located in the corridor or space adjacent to the stairway or pipe shaft enclosure and connected through the wall to the standpipe.

17-22 A fire standpipe system which has been out of service for 4 years, due to no occupancy of the building, can be restored to service, according to Code, by testing it with air at a maximum pressure of _____ psi.

(A) 25

(B) 50

(C) 75

(D) 100

NFPA 14

Answer: A

Code response: When a standpipe system has been out of service for a number of years, before it is filled with water and restored to service, it shall be tested with air at a pressure not exceeding 25 psi to determine its tightness.

17-23 **When preparing a fire standpipe system for tests and inspection, according to Code, piping between the siamese connection and the check valve must be:**

(A) tested by air pressure

(B) smoke tested

(C) tested with a mixture of peppermint and oil

(D) tested hydrostatically

NFPA 14

Answer: D

Code response: Piping between the check valve and the fire department connection shall be tested hydrostatically.

17-24 **The word "should," when used in the Code, means:**

(A) a mandatory requirement

(B) advised but not required

(C) the value may be approximate

(D) equivalent value

NFPA 14

Answer: B

Code response: The word "should" indicates a recommendation but is not required.

17-25 **Standpipe systems are grouped into three general classes of service in extinguishing fires. Those systems in the category not regulated by the jurisdictional authority and not meeting Code standards, must be clearly marked with a sign that states:**

(A) "for fire department use only."

(B) "for those trained in handling heavy fire streams."

(C) "for use primarily by building occupants."

(D) "for fire brigade use only."

NFPA 14

Answer: D

Code response: Buildings equipped with fire standpipe and hose systems but not meeting standard requirements of the Code, nor required to be under the authority having jurisdiction, shall be marked "For Fire Brigade Use Only."

17-26 **A fire standpipe having an inadequate supply of water but enough to keep the piping full, is classified by Code as a:**

(A) dry standpipe

(B) wet standpipe

(C) inadequate wet standpipe

(D) adequate wet standpipe

NFPA 14

Answer: A

Code response: A fire standpipe having a small water supply, but enough to keep the piping full, shall be considered a dry standpipe.

17-27 When designing a fire standpipe system, the size of standpipes, according to Code, is governed by:

(A) the length of the building
(B) the type of materials used in the construction of the building
(C) the number of floors in the building
(D) the width of the building

NFPA 14

Answer: C

Code response: The building height governs the size of standpipes in a given case.

17-28 The minimum size of fire standpipes for a building not exceeding 100 feet in height is _____ inches.

(A) 3
(B) 4
(C) 5
(D) 6

NFPA 14

Answer: B

Code response: The fire standpipes in a building which does not exceed 100 feet in height shall be no less than 4 inches.

17-29 An office building no higher than 12 stories is required by Code to have each fire standpipe sized to provide a minimum flow of _____ gpm.

(A) 100
(B) 150
(C) 200
(D) 250

NFPA 14

Answer: A

Code response: Fire standpipe systems for Class II (office buildings) shall be sized so that each standpipe will provide a minimum flow of 100 gpm.

17-30 When installing 6-inch steel pipe fire lines underground, the Code requires that they be protected against corrosion:

(A) by painting pipe and fittings with an approved corrosion-resistive coating
(B) by using only galvanized steel pipe
(C) by wrapping pipe and fitting with an approved material
(D) before they are buried

NFPA 14

Answer: D

Code response: If steel pipe is used underground, where corrosive conditions exist, it shall be protected against corrosion before being buried.

17-31 When designing a wet standpipe system to meet Code specifications for a building requiring three fire standpipes, such standpipes must:

(A) have the underground piping sized not less than 6 inches

(B) be a minimum of 4 inches in diameter

(C) be interconnected at the bottom

(D) be interconnected at the bottom and at the top

NFPA 14

Answer: C

Code response: In a building requiring two or more standpipes, they shall be interconnected at the bottom.

17-32 According to Code, a textile manufacturing building having a combined fire protection system with partial automatic sprinkler protection must have a water supply increased by _____ gpm, or equal to the hydraulically calculated sprinkler demand.

(A) 150

(B) 200

(C) 350

(D) 500

NFPA 14

Answer: D

Code response: In buildings classified for light hazard occupancies having a combined system with partial automatic sprinkler protection, the water supply requirements shall be increased by an amount equal to the hydraulically calculated sprinkler demand, or by 500 gpm.

17-33 Theaters and auditoriums, excluding stages and prosceniums, having combined systems which are fully sprinklered may, according to Code:

(A) omit using a 1½-inch hose for the building occupants

(B) provide a 1½-inch hose for use by the building occupants

(C) not require full size standpipes

(D) have standpipes not less than 4 inches in diameter

NFPA 14

Answer: A

Code response: In Class II service for buildings completely sprinklered, having combined fire protection systems, you may omit the 1½-inch hose usually required for the building occupants.

17-34 When installing a fire protection system with welded joints, the piping material, according to Code, must withstand pressures up to _____ psi.

(A) 175

(B) 250

(C) 300

(D) 375

NFPA 14

Answer: C

Code response: When welded pipe is used and joined by welding, the minimum wall thickness shall withstand pressures up to 300 psi.

17-35 Code specifies that when a fire protection system is connected to a public water system, it must be:

(A) protected from backflow by a minimum 4-inch check valve

(B) controlled by indicator post valves

(C) controlled by at least 2 gate valves

(D) provided with a water flow alarm fitting

NFPA 14

Answer: B

Code response: Fire protection systems, when connected to public water systems, shall be controlled by indicator post valves.

17-36 When installing fire standpipe risers, sometimes the horizontal runs must unavoidably pass through exposed areas subject to drastic temperature changes. When this situation exists, the Code requires that:

(A) the pipe be graded to drain dry

(B) the piping material be constructed of malleable iron only

(C) safeguards be provided to prevent freezing

(D) safeguards be provided to prevent water in pipe from overheating

---

NFPA 14

Answer: C

Code response: When fire standpipe risers, including horizontal runs, pass through unheated areas, safeguards shall be provided to prevent freezing.

17-37 Where water pressures range from 175 to 300 psi, according to Code, the _____ standards will permit the use of standard wall pipe with extra heavy valves.

(A) ANSI

(B) ASTM

(C) AUTOSPKR

(D) NFPA

NFPA 14

Answer: A

Code response: In installations where water pressures range from a minimum of 175 to a maximum of 300 psi, the ANSI standards permit the use of standard wall pipe and extra heavy valves.

17-38 Fire standpipe systems, according to Code, are grouped into three general classes of service: Class I, Class II, and Class III. For a Class I service, the statement that is most nearly correct is:

(A) Class I service is primarily installed for use by fire departments

(B) Class I service is generally installed for use by the building occupants

(C) Class I service is installed for use by either fire departments or those properly trained in heavy hose stream use

(D) Class I service is designed and installed for fire brigade use only

---

NFPA 14

Answer: A

Code response: Fire standpipe systems for Class I service are designed primarily for use by fire departments.

17-39 The 2½-inch piping from fire standpipe riser to hose cabinet, according to Code, must be properly supported if the developed length exceeds _____ inches.

(A) 12
(B) 18

(C) 24
(D) 36

NFPA 14

Answer: B

Code response: The horizontal runs from the fire standpipe to the hose valve and cabinet over 18 inches in length shall be provided with hangers.

17-40 According to Code, the hangers used to support vertical standpipes must be able to uphold a water-filled pipe, as well as an additional _____ pounds at point of hanging.

(A) 100
(B) 150

(C) 200
(D) 250

NFPA 14

Answer: D

Code response: Fire standpipes shall be substantially supported from the building structure which must support the added load of the water-filled pipe, plus an additional 250 pounds at the point of hanging.

17-41 According to Code, standpipe risers must be properly supported:

(A) to prevent movement upward
(B) by attachments directly to the risers

(C) at every other floor
(D) by clamps with heavy-duty set screws

NFPA 14

Answer: B

Code response: Fire standpipe risers shall be supported by approved attachments directly to each riser.

17-42 Code mandates that 4-inch fire standpipes installed horizontally be supported with approved hangers spaced no further than _____ feet apart.

(A) 8
(B) 10

(C) 12
(D) 15

NFPA 14

Answer: D

Code response: A maximum distance of 15 feet between hangers shall be provided on horizontal fire standpipe runs.

17-43 Fire hose cabinets installed in a **14**-story apartment building, according to Code, must be located no more than _____ feet from the finished floor.

(A) 4

(B) 5

(C) 6

(D) 7

NFPA 14

Answer: C

Code response: Hose connections in no case shall be over 6 feet from the floor.

17-44 Approved dry standpipes, according to Code, shall not be:

(A) installed in stair enclosures

(B) concealed in building walls

(C) installed in pipe shafts

(D) located in stairway platforms constructed of concrete and steel

NFPA 14

Answer: B

Code response: When approved for use by local authority having jurisdiction, dry fire standpipes shall not be concealed in building walls.

17-45 Code mandates that fire standpipes in a 6-story building be at least _____ inches in diameter.

(A) 2½

(B) 3

(C) 4½

(D) 5

NFPA 14

Answer: A

Code response: Fire standpipes exceeding 50 feet in height shall be no less than 2½ inches in diameter.

Note: Most local codes require fire standpipes to be a minimum of 4 inches in diameter for buildings up to 75 feet high.

17-46 Fire hose stations provided for nursing homes and used by building occupants shall, according to Code, be equipped with listed 1½-inch lined fire hoses not more than _____ feet long.

(A) 50

(B) 75

(C) 100

(D) 125

NFPA 14

Answer: C

Code response: Fire hose stations for use by building occupants shall be equipped with not more than 100 feet of listed 1½-inch lined fire hoses.

Q **17-47 According to Code, each zone in a fire protection system of a high-rise building where two or more zones are used:**

(A) should have hose connections on the street side of the building

(B) must have an approved automatic drip installed

(C) must be provided a fire department connection

(D) must have an approved indicating-type valve located outside at a safe distance from the building

---
NFPA 14

Answer: C

Code response: Fire protection systems in high-rise buildings having two or more zones shall have a fire department connection provided for each zone.

Q **17-48 The fire standpipe system of a 30-story building is equipped with gravity tanks. According to Code, the gravity tanks should connect:**

(A) at the base of the fire standpipe system

(B) to the fire standpipe system at the 10th floor level

(C) at the top of the fire standpipe system

(D) to the fire standpipe system at the 29th floor level

NFPA 14

Answer: C

Code response: Connections from gravity tanks shall be made to the top of the standpipe system.

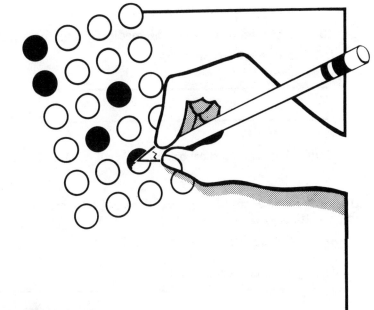

Chapter 18

Nonflammable Medical Gas Systems

Your local plumbing code regulates the types of materials, installation practice and testing methods for nonflammable medical gas piping in hospitals and similar buildings. Nonflammable medical gases include oxygen, nitrous oxide, medical compressed air, carbon dioxide, helium, nitrogen, and mixtures of those gases when used for medical purposes.

Nonflammable medical gas systems are only a small part of a plumber's job. In fact, some plumbers may work an entire career without installing a nonflammable gas system. But a plumbing contractor's certification qualifies the contractor or the firm to do this type of work. Therefore, state and some local examinations may include several questions on nonflammable medical gas.

This short chapter should cover about all you'll need to know about this specialized topic.

Q 18-1 In accordance with Code, medical gas piping is not to be installed in a trench with:

(A) water service piping
(B) fuel oil lines

(C) building sewer
(D) storm water drain

NFPA 56F

Answer: B

Code response: Medical gas piping shall not be installed where it may be exposed to contact with oil.

Q 18-2 Medical gas piping, according to Code, must be properly sized to deliver the maximum volumes specified by using:

(A) only specified type pipe and fittings

(B) good engineering practice

(C) piping and fittings one pipe size larger than the maximum volume required
(D) the cryogenic liquid cylinder capacity

NFPA 56F

Answer: B

Code response: Medical gas piping shall be sized to deliver maximum volume specified in conformity with good engineering practice.

Q 18-3 When installing 1¼-inch horizontal brass piping to deliver the maximum volumes of medical gas as specified by Code, the piping must be supported at distances not to exceed _____ feet.

(A) 4
(B) 6

(C) 8
(D) 10

NFPA 56F

Answer: D

Code response: Horizontal piping 1¼ inches and larger shall be supported at distances not to exceed 10 feet.

Q 18-4 When repairing station outlets for low pressure medical gases, according to Code, "o" rings:

(A) may be used

(B) may not be used

(C) shall be constructed of a special approved material
(D) may be used if in accordance with Compressed Gas Association specifications

---

NFPA 56F

Answer: A

Code response: For replacement purposes, common parts such as "o" rings are permissible.

18-5 To prevent physical damage to equipment, Code mandates that medical gas station outlets in patient rooms be:

(A) located near the door, entering the room

(B) installed above the head of the bed

(C) located at an appropriate height above the floor

(D) recessed in the wall only

NFPA 56F

Answer: C

Code response: When installing medical gas station outlets in patient rooms, the outlets shall be located at an appropriate height above the floor to prevent physical damage to equipment connected to the outlet.

18-6 After installation of all piping and station outlets, and before the walls are closed, Code requires that each section be tested with a minimum pressure of _____ psig.

(A) 50

(B) 75

(C) 125

(D) 150

NFPA 56F

Answer: D

Code response: After installation of gas piping and outlets and before closing the walls, each section shall be subjected to a minimum pressure of 150 psig.

18-7 All components of a completed medical gas piping system must be closed tight and filled with test gas for a minimum period of 24 hours. According to Code, the test gas must be comprised of:

(A) nitrous oxide

(B) helium

(C) oil-free dry air

(D) compressed air

NFPA 56F

Answer: C

Code response: The completely assembled station outlets and all components shall be subjected to a 24-hour standing pressure test. The **test gas** shall be oil-free dry air.

18-8 You and your crew have completed installation of the medical gas piping system. Before installing the system components, Code requires that all lines be blown clear by using:

(A) oil-free dry nitrogen

(B) medical compressed air

(C) oxygen

(D) air

NFPA 56F

Answer: A

Code response: Before the installation of system components, the piping system shall be blown clear by means of oil-free dry nitrogen.

Q 18-9 In accordance with Code, the plumber must be certain that all test gas is removed from any newly-installed medical gas piping system by making sure the gas:

(A) flows through and out each outlet until it has no blue haze left

(B) flows through and out each outlet until a flame will no longer ignite the gas

(C) no longer discolors a white cloth which it impinges on

(D) flows freely from each outlet giving off no odor

NFPA 56F

Answer: C

Code response: Purge gas shall be allowed to impinge upon a white cloth until no evidence of discoloration is evident.

Q 18-10 The gas content of a medical gas piping system must be readily identifiable by labeling at intervals not to exceed _____ feet, according to Code.

(A) 8

(B) 10

(C) 15

(D) 20

NFPA 56F

Answer: D

Code response: Appropriate labeling with the name of the gas contained shall appear on the piping at intervals of not more than 20 feet.

Q 18-11 The Code requires that medical gas piping not be supported by:

(A) other piping

(B) pipe hooks

(C) metal pipe straps

(D) bands

NFPA 56F

Answer: A

Code response: Medical gas piping shall not be supported by other piping.

Q 18-12 According to Code, the list of nonflammable medical gases does not include:

(A) helium

(B) carbon dioxide

(C) nitrogen

(D) ethylene

NFPA 56F

Answer: D

Code response: Nonflammable medical gases include oxygen, nitrous oxide, medical compressed air, carbon dioxide, helium, nitrogen and mixtures of such gases.

18-13 Code requires that vertically installed 1¼-inch copper pipe conveying medical gas be supported at:

(A) 8-foot intervals
(B) 10-foot intervals

(C) every floor level
(D) every other floor level

NFPA 56F

Answer: C

Code response: Vertically installed medical gas piping of 1¼ inches in diameter shall be supported at every floor level.

18-14 Before erection, medical gas systems, according to Code, must be thoroughly cleaned in:

(A) a hot solution of sodium carbonate
(B) an organic solvent only

(C) kerosene
(D) gasoline

NFPA 56F

Answer: A

Code response: Before erection, piping, valves and fittings shall be cleaned by washing in a hot solution of sodium carbonate.

18-15 Piping, valves and fittings for oxygen service, according to Code,

(A) shall be cleaned by washing in a hot solution of sodium carbonate
(B) shall be cleaned by washing in a hot solution of trisodium phosphate

(C) need not be cleaned before installation
(D) should be thoroughly cleaned with a solution of kerosene or gasoline

NFPA 56F

Answer: C

Code response: Before erection, piping, valves and fittings for oxygen service need not be thoroughly cleaned.

18-16 According to Code, ¾-inch horizontally installed seamless type K copper tubing conveying medical gas must be supported at intervals of not more than _____ feet.

(A) 4
(B) 6

(C) 8
(D) 10

NFPA 56F

Answer: C

Code response: ¾″ horizontal copper tubing shall be supported at intervals not to exceed 8 feet.

Q 18-17 Nonflammable medical gases, according to Code, include:

(A) nitrous oxide
(B) ethylene

(C) ethyl chloride
(D) cyclopropane

NFPA 56F

Answer: A

Code response: Nonflammable medical gases include nitrous oxide.

Q 18-18 Air compressors used for medical gas supply systems, according to Code, have to be:

(A) single, piston-type, oil lubricated
(B) duplex, piston-type, oil lubricated

(C) single, oil-free units
(D) duplex, oil-free units

NFPA 56F

Answer: D

Code response: Air compressors shall consist of two or more oil-free duplexed units.

Q 18-19 Code mandates that a nitrogen supply system for a medical gas system be capable of delivering at least ____ psig to all outlets at maximum flow.

(A) 100
(B) 130

(C) 160
(D) 190

NFPA 56F

Answer: C

Code response: A nitrogen system shall be capable of delivering at least 160 psig to all outlets at maximum flow.

Q 18-20 In medical gas supply systems, according to Code, a check valve must be installed between the ____ to prevent the loss of gas in the event a cylinder lead (pigtail) fails.

(A) cylinder lead and the manifold header

(B) cylinder and the pressure relief valve

(C) main shut off valve and the pressure regulator
(D) line regulator and the main supply pipe

NFPA 56F

Answer: A

Code response: A check valve shall be installed between the cylinder lead and the manifold header.

Q 18-21 According to Code, if natural venting is used for the location of a medical gas supply system whose capacity is 2,200 cubic feet (connected and in storage), the minimum vent opening in free area must be ____ square feet.

(A) 0.50

(B) 0.72

(C) 0.78

(D) 0.82

NFPA 56F

Answer: A

Code response: The vent opening for any system whose capacity is more than 2,000 cu. ft. shall be a minimum of 72 square inches in total free area. It must be vented to the outside.

Solution: 144 sq. in. = 1 sq. ft.

$$\frac{72 \text{ sq. in.}}{144 \text{ sq. in.}} = 0.50 \text{ sq. ft. (or 1/2 of 1 sq. ft.)}$$

Q 18-22 According to Code, medical compressed air is defined as air which is:

(A) supplied by an air compressor serving only medical patient facilities

(B) compressed to 30 psig

(C) reconstituted from oxygen U.S.P. and nitrogen N.F.

(D) supplied by an air compressor approved for medical gas

NFPA 56F

Answer: C

Code response: Medical compressed air is defined as air which has been reconstituted from oxygen U.S.P. and nitrogen N.F.

Part Four
Plumbing Isometrics, Piping Diagrams and Plumbers' Math

- Isometric Drawings
- Basic Plumbing Mathematics

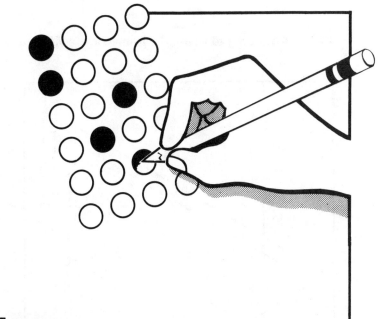

Chapter 19

Isometric Drawings

Isometric drawings are the means of communication among plumbing professionals. They are to the plumber what sheet music is to the musician. If you can't read an isometric, you can't understand what's required in a DWV (drainage, waste and vent) or piping system.

Lines on an isometric represent pipe and fittings. They're used by the plumbing contractor to estimate the cost of new work and to show the job foreman how to rough-in a particular job. All experienced, professional plumbers should be able to make and interpret isometric drawings.

In the early 1980's several states began to include in their exams questions that required some understanding of isometric drawings. These questions weren't particularly demanding. In fact, I assumed that most applicants would have little trouble with this part of the test. Unfortunately, I was wrong. When the test scores were in, we discovered that most examinees knew relatively little about isometric drawings. The percentage of those that passed the first Florida exam with isometrics dropped to 10% below normal.

A plumber who can't read isometric drawings is operating under a major handicap. Don't make that mistake. Don't take the plumbing exam until you're comfortable with isometric drawings.

Some exams will give you a "zoned" drawing that shows part of a DWV system. But most ques-tions will show a "mini" drawing. This tests your ability to interpret and correct isometric drawings.

Here are the directions for the sample questions in this chapter:

If the drawing is *correct*, you are to circle the letter "A". If the drawing is *incorrect*, you are to circle the letter "B". For drawings that are incorrect or not complete, you are to draw them *correctly* beside the incorrect figure. Corrected drawings are shown at the end of this chapter.

Following is a list of the abbreviations used in the isometrics in this chapter:

P-1	single compartment sink
P-2	double compartment sink
P-3	floor drain
P-4	floor sink
P-5	garbage can wash
P-6	food waste disposal
P-7	bucket type floor drain
LAV	lavatory
WC	water closet
TUB	bathtub
CO	cleanout
VTR	vent through roof
GW	greasy waste
BD	building drain
CFH	cubic feet/hour

Q 19-1 According to Code, the isometric drawing shown below is _____. Pipe size is __not__ a factor.

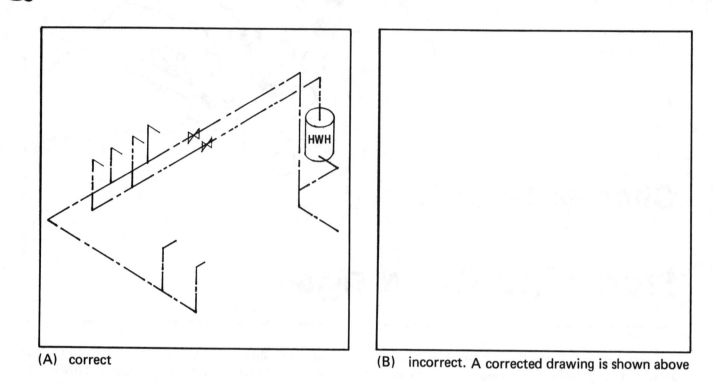

(A) correct   (B) incorrect. A corrected drawing is shown above

Q 19-2 According to Code, the isometric drawing shown below is _____. Pipe size is __not__ a factor.

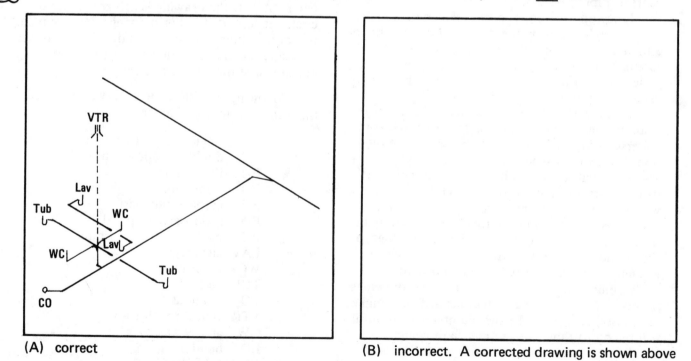

(A) correct   (B) incorrect. A corrected drawing is shown above

Note: Answers to questions on this page are given at the end of the chapter.

Q 19-3   According to Code, the drawing shown below is _____ . Pipe size is **not** a factor.

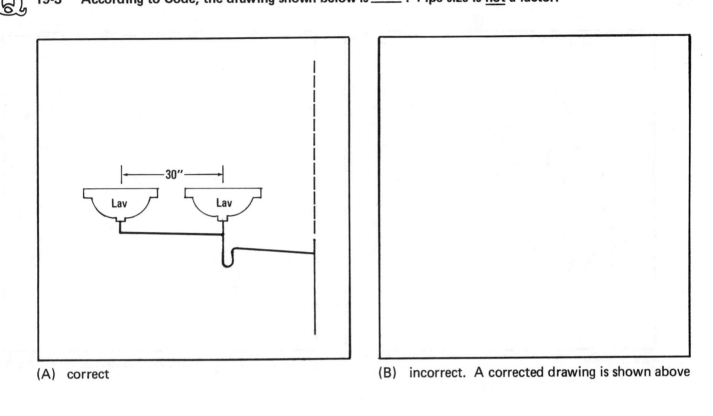

(A)   correct

(B)   incorrect. A corrected drawing is shown above

Q 19-4   According to Code, the isometric drawing shown below is _____ . Pipe size is **not** a factor.

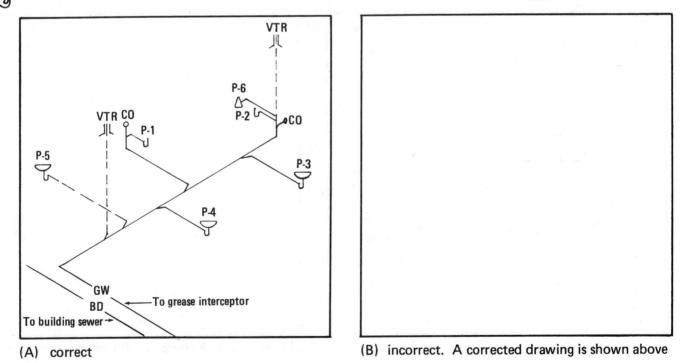

(A)   correct

(B)   incorrect. A corrected drawing is shown above

Note: Answers to questions on this page are given at the end of the chapter.

19-5 According to Code, the isometric drawing shown below is _____ . Pipe size _is_ a factor.

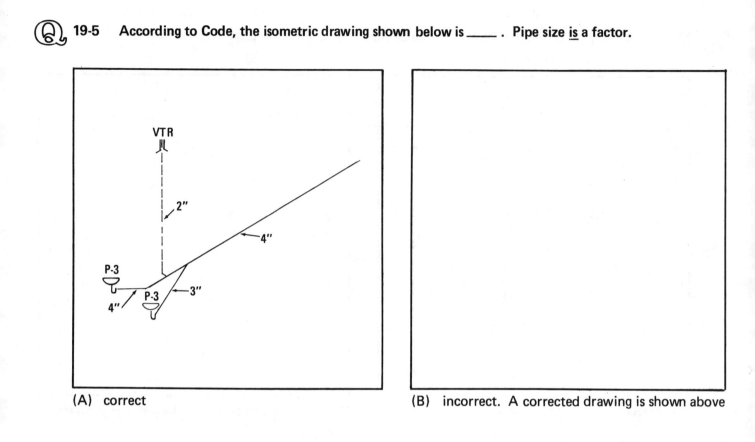

(A) correct

(B) incorrect. A corrected drawing is shown above

19-6 According to Code, the drawing shown below is _____ .

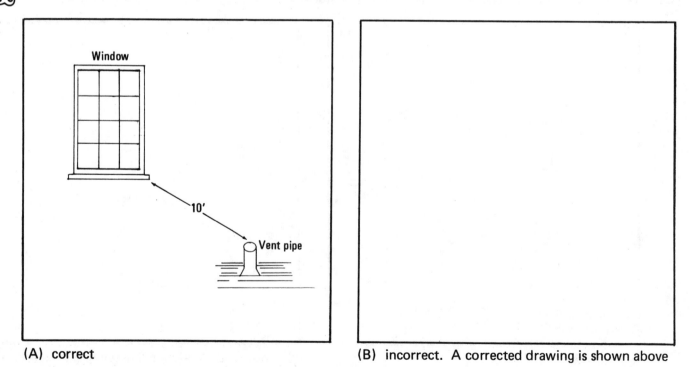

(A) correct

(B) incorrect. A corrected drawing is shown above

Note: Answers to questions on this page are given at the end of the chapter.

Q 19-7 **According to Code, the isometric drawing shown below is _____ . Pipe size _is_ a factor.**

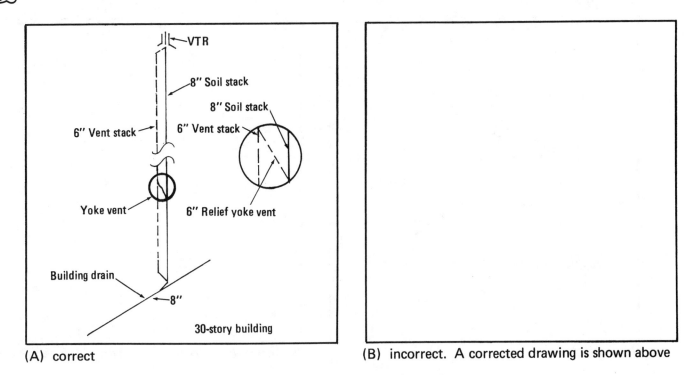

(A) correct

(B) incorrect. A corrected drawing is shown above

Q 19-8 **According to Code, the isometric drawing shown below is _____ . Pipe size _is_ a factor.**

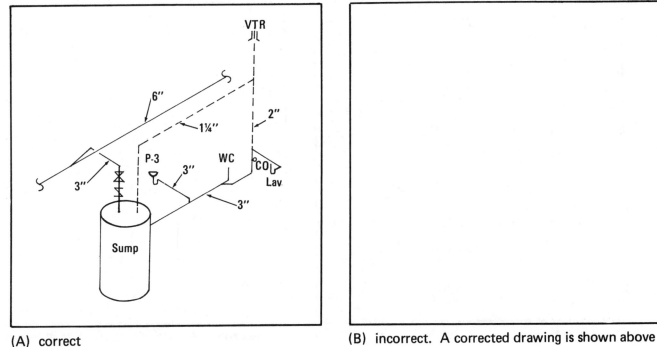

(A) correct

(B) incorrect. A corrected drawing is shown above

Note: Answers to questions on this page are given at the end of the chapter.

Q 19-9 According to Code, the isometric drawing shown below is _____ . Pipe size and oil interceptor size are <u>not</u> factors.

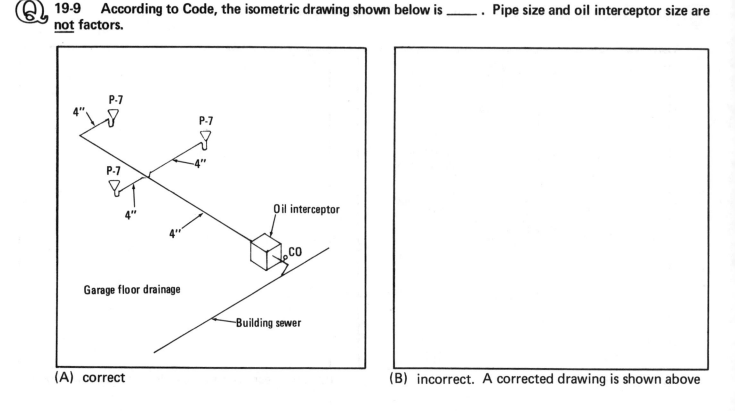

(A) correct

(B) incorrect. A corrected drawing is shown above

Q 19-10 According to Code, the isometric drawing shown below is _____ . Pipe size <u>is</u> a factor.

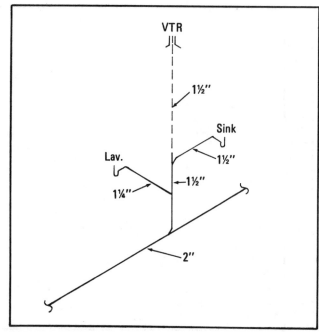

(A) correct

(B) incorrect. A corrected drawing is shown above

Note: Answers to questions on this page are given at the end of the chapter.

19-11 According to Code, the isometric drawing shown below is ____ . Pipe size is **not** a factor.

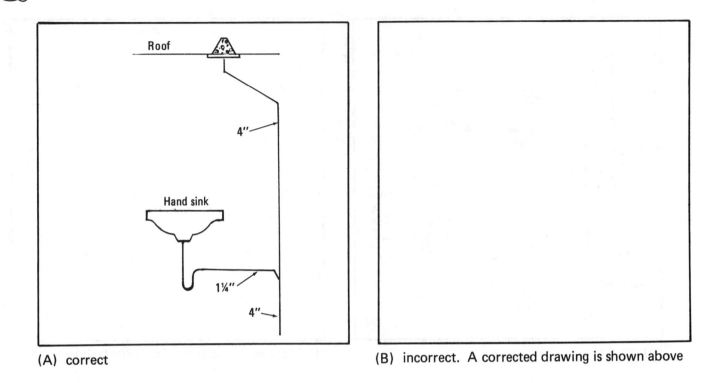

(A) correct   (B) incorrect. A corrected drawing is shown above

19-12 According to Code, the drawing shown below is ____ . Pipe size is **not** a factor.

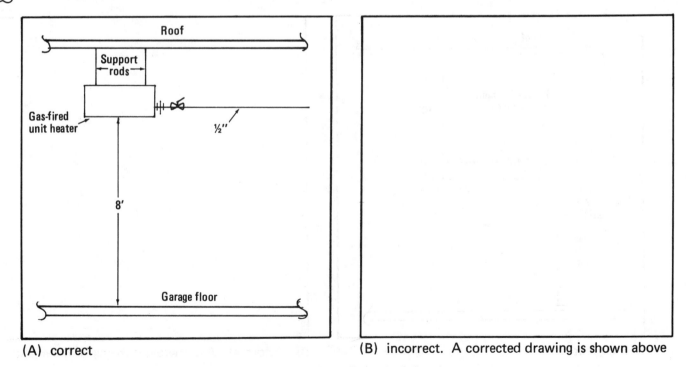

(A) correct   (B) incorrect. A corrected drawing is shown above

Note: Answers to questions on this page are given at the end of the chapter.

Q **19-13** **According to Code, the piping diagram shown below is _____ . Pipe size and length are not factors.**

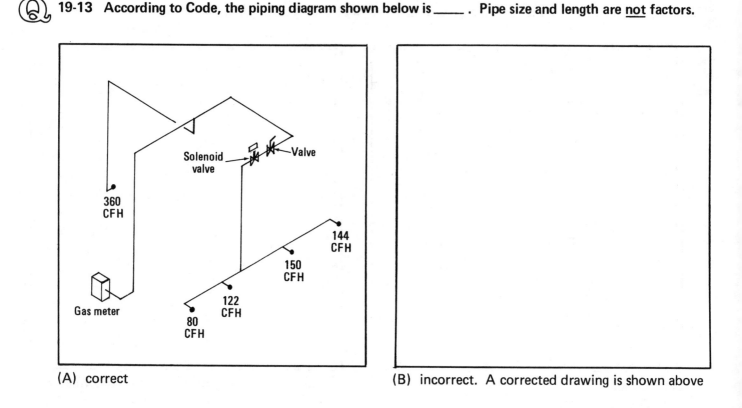

(A) correct

(B) incorrect. A corrected drawing is shown above

Q **19-14** **According to Code, the drawing as shown below is _____ .**

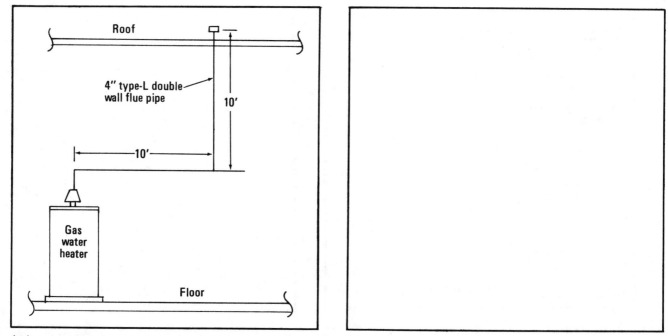

(A) correct

(B) incorrect. A corrected drawing is shown above

Note: Answers to questions on this page are given at the end of the chapter.

19-15 **According to Code, the drawing as shown below is:**

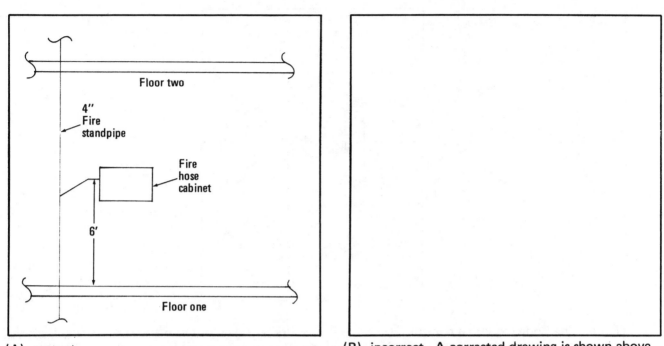

Floor two

4"
Fire
standpipe

Fire
hose
cabinet

6'

Floor one

(A) correct

(B) incorrect. A corrected drawing is shown above

Note: Answers to questions on this page are given at the end of the chapter.

Answers

19-1   **Answer: (B) Drawing is incorrect. A corrected drawing is shown below:**

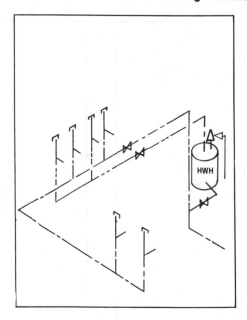

Note: The Code requires a **control valve** and a **P and T relief valve** for hot water heaters, and that the piping system for this kind of installation be protected with air chambers.

19-2   **Answer: (A) Drawing is correct.**

19-3   **Answer: (A) Drawing is correct.**

19-4   **Answer: (B) Drawing is incorrect. A corrected drawing is shown below:**

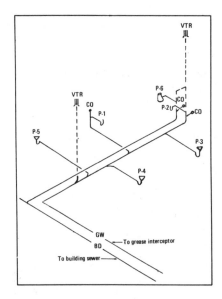

Note: The Code prohibits a food waste disposal unit from discharging into any grease interceptor.

Answers

19-5   Answer: (A) Drawing is correct.

19-6   Answer: (A) Drawing is correct.

19-7   Answer: (A) Drawing is correct.

19-8   Answer: (B) Drawing is incorrect. A corrected drawing is shown below:

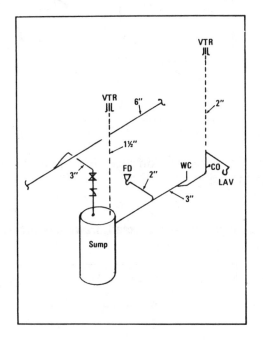

Note Code requires that sump vents shall in no case be smaller than 1½ inches, and that they must extend separately through the roof.

19-9   Answer: (A) Drawing is correct.

Answers

19-10 Answer: (B) Drawing is incorrect. A corrected drawing is shown below:

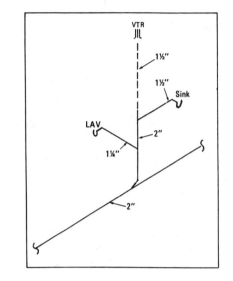

Note: Code requires that each wet-vented section shall be a minimum of one pipe size larger than the required minimum waste pipe of the upper fixture.

19-11 Answer: (B) Drawing is incorrect. A corrected drawing is shown below:

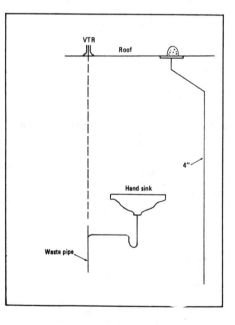

Note: Code requires that all plumbing fixtures used to receive or discharge liquid wastes shall be connected properly to the drainage system, and that rainwater piping shall not be used as waste pipes.

Answers

19-12 Answer: (A) Drawing is correct.

19-13 Answer: (B) Drawing is incorrect. A corrected drawing is shown below:

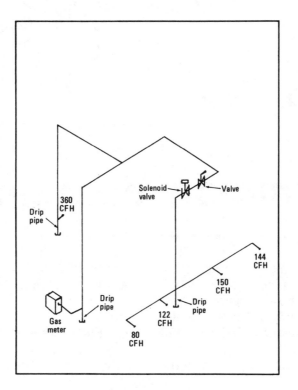

Note: Code requires that appropriate drip pipes be provided to receive condensation that may form within the pipe, and that gas branch pipes connecting to horizontal pipes connect at the top or the side, and never from the bottom.

19-14 Answer: (A) Drawing is correct.

19-15 Answer: (A) Drawing is correct.

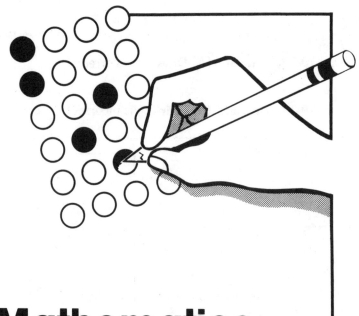

Chapter 20

Basic Plumbing Mathematics

Every plumber needs to make mathematical calculations. For example, you need to know how to measure sections of pipe, how to figure the area of a tank, and how to do simple hydraulic calculations involving volume, pressures, velocity and weight of water.

This chapter tests your knowledge of only the basic mathematics — the type that's most likely to be on the plumber's exam. But these questions cover a broad spectrum of the most typical exam problems.

Each question has only one solution. But there may be several ways to arrive at that answer. The notes show an explanation if it's necessary.

Q 20-1 If 600 feet of pipe weighs 900 pounds, 140 feet of like pipe would weigh ____ pounds. Select the closest answer.

(A) 180

(B) 195

(C) 211

(D) 217

Answer: C

Solution: $\dfrac{900 \text{ lb.}}{600 \text{ ft.}} \times 140' =$

 1.5 lb. $\times$ 140' = 210 lb.

Note: Divide total pounds by length of pipe to arrive at pounds per foot of pipe. Then multiply pounds per foot times the number of feet of pipe to arrive at total pounds.

Q 20-2 A 60 gallon water heater weighs 52.5 pounds when empty and ____ pounds when full. Select the closest answer.

(A) 396

(B) 501

(C) 553

(D) 585

Answer: C

Solution: 60 gal. x 8.3453 + 52.5 = 553.22 lb.

Note: Multiply total gallons by 8.3453 (lb. per gal.). Then add 52.5 (weight of empty tank) to arrive at total weight of water heater when full.

Q 20-3 In accordance with Code, find the minimum drop per 100 feet of a horizontal building sewer 6 inches in diameter. Select the closest answer.

(A) 8.75 inches

(B) 12.50 inches

(C) 15.25 inches

(D) 17.15 inches

Answer: B

Solution: 1/8" per ft. x 100' = 12½"

Note: The Standard Plumbing Code and the Uniform Plumbing Code require a minimum drop of 1/8" per foot for horizontal drainage piping sized 6" in diameter. (To find a decimal answer, use decimal equivalent 0.125" in lieu of 1/8".)

20-4 Given: A roof that is **100 feet long, 50 feet wide, with a parapet wall 6 inches high. The roof leaders become plugged and unable to drain. Under such conditions the roof area will fill up with a volume of** _____ **gallons of rain water. Select the closest answer.**

(A) 18,700

(B) 18,950

(C) 19,375

(D) 24,400

Answer: A

Solution: 100' x 50' x ½ ft (.5) x 7.48 gal./cu. ft. = 18,700 gal.

20-5 If the inside dimensions of a grease interceptor are **9'0" long, 4'6" wide, 5'0" deep, with a 20% freeboard, the liquid capacity of the interceptor is** _____ **gallons. Select the closest answer.**

(A) 1,150.25

(B) 1,211.76

(C) 1,430.40

(D) 1,565.00

Answer: B

Solution: 9' x 4.5' x 5' = 202.5 cu. ft. (gross)

202.5 cu. ft. (gross) — .20 freeboard (40.5 cu. ft.) = 162 cu. ft. (net) x 7.48 gal/cu. ft. = 1211.76 gal.

Note: Be sure to change 6" (in the width measurement) into 1/2' (decimal equivalent, .5') before multiplying.

20-6 A 4-inch soil stack is approximately **60 feet high. If all openings were plugged and the stack filled with water, the pressure at base of stack would equal** _____ **pounds per square inch. Select the closest answer.**

(A) 21.07

(B) 23.25

(C) 24.12

(D) 26.01

Answer: D

Solution: 60' x 0.4335 = 26.01 psi

Note: To find the amount of pressure (lb/sq.in) of a column of water, multiply the height of the column in feet times 0.4335. (This figure is found in an appropriate reference table.)

20-7 A tank **32 inches long, 10 inches high and 8 inches wide would hold** _____ **gallons of water. Select the closest answer.**

(A) 10.04

(B) 11.08

(C) 12.20

(D) 12.50

Answer: B

Solution: 32" x 10" x 8" ÷ 231 cu.in./gal. = 11.08 gal.

Q **20-8** Given: The architect specifies sheet copper safe pans for all showers within a building. Inside dimensions of shower stalls are 3' x 4' with a 6" turnup. According to Code, each shower safe pan would weigh _____ pounds.

(A) 12   (C) 14

(B) 13   (D) 15

Answer: D

Solution: 4' x 5' x .75 lb./sq. ft. = 15 lb

Note: Most Codes require copper safe pans to weigh no less than 12 oz. per sq. ft., which is ¾ lb. or .75 lb./sq. ft.

Q **20-9** Find, in accordance with Code, the maximum drop per 110 feet of a horizontal building drain which is 4 inches in diameter. Select the closest answer.

(A) 30"   (C) 55"

(B) 45"   (D) 65"

Answer: C

Solution: ½" (.5") x 110' = 55"

Note: Most Codes accept a maximum ½" drop per foot for horizontal drains.

Q **20-10** Given: A roof that is 100 feet long, 50 feet wide, with a parapet wall 6 inches high. If roof leaders should become clogged and fill the roof to overflowing, the weight of the rain water on the roof would be most nearly _____ pounds.

(A) 150,700   (C) 180,250

(B) 156,250   (D) 185,700

Answer: B

Solution: 100' x 50' x .5' (6") x 62.5 lb./cu. ft. = 156,250 pounds

20-11 You are cutting four pieces of pipe end-to-end, with the following dimensions: 3'2-1/4", 4'7-3/8", 5'4-5/8", 6'6-7/8". To make these cuts, you would need ____ of pipe. Select the closest answer.

(A) 19'6"

(B) 19'10-1/8"

(C) 19'-9"

(D) 20'2-3/16"

Answer: C

Solution: Change all feet into inches. Also change inches into decimals where appropriate.

Fraction	=	Decimal
1/4	=	.250
3/8	=	.375
5/8	=	.625
7/8	=	.875

3'2-1/4" = (3 x 12" = 36" + 2" + .25") = 38.250"

4'7-3/8" = (4 x 12" = 48" + 7" + .375") = 55.375"

5'4-5/8" = (5 x 12" = .60" + 4" + .625") = 64.625"

6'6-7/8" = (6 x 12" = 72" + 6" + .875") = 78.875"

Total inches 237.125"

$$\frac{237.125"}{12"/ft.} = 19.76', \text{ or approximately } 19\tfrac{3}{4}' \text{ which is } 19'9"$$

20-12 A rectangular tank is filled with water to a height of 5'0". The pressure exerted on the bottom of the tank is most nearly ____ psi.

(A) 1.98

(B) 2.17

(C) 2.46

(D) 2.65

Answer: B

Solution: 5' x .4335 = 2.1675, or 2.17 psi

Note: To find the amount of pressure (lb/sq.in.) of a column of water, multiply the height of the column in feet by 0.4335. (This figure is found in an appropriate reference table.)

20-13 A septic tank's inside measurements are 8'6" long, 4'2" wide, 4'8" deep, with an 8" air space. The septic tank would have a liquid capacity of ____ gallons. Select the closest answer.

(A) 900

(B) 975

(C) 1,030

(D) 1,060

Answer: D

Solution: Change inches into decimals where appropriate.

(8'6" = 8.5' 4'2" = 4.167', 4'0" = 4'.) Then multiply 8.5' x 4.167' x 4.0' x 7.48 gal./cu.ft. = 1059.75144 gal.

Note: Air space must be subtracted from the depth before calculating liquid capacity.

Q 20-14 According to any hydrostatic table, a gallon of water weighs approximately _____ pounds.

(A) 2.31　　　　　　　　　　　　　　　(C) 3.61
(B) 8.34　　　　　　　　　　　　　　　(D) 7.47

Answer: B

Solution: A gallon of water (United States Standards) weighs 8-1/3 pounds.

Q 20-15 Offsets in water piping can be made with a number of acceptable fittings. If the constant 2.613 is used to compute the developed length of the offset piping, then the fittings to use to make the offset must be:

(A) 5-5/8⁰　　　　　　　　　　　　　　(C) 22-1/2⁰
(B) 11-1/4⁰　　　　　　　　　　　　　　(D) 30⁰

Answer: C

Note: The constant 2.613 x offset equals the diagonal (piece of pipe) needed to join two 22-1/2⁰ ells. Refer to your math or reference book for complete constant data.

Q 20-16 150 cubic feet of water will weigh most nearly _____ pounds. Select the closest answer.

(A) 8,750　　　　　　　　　　　　　　(C) 9,632
(B) 9,375　　　　　　　　　　　　　　(D) 9,813

Answer: B

Solution: 150 ft.^{3} x 62.5 lb/cu. ft. = 9,375 lb.

Q 20-17 A PVC fitting that turns an angle of 22½ degrees is a _____ bend.

(A) 1/4　　　　　　　　　　　　　　　(C) 1/8
(B) 1/5　　　　　　　　　　　　　　　(D) 1/16

Answer: D

Solution: $\dfrac{22.5^0}{360^0}$ = .0625, which is equivalent to a 1/16-bend

Note: To find a fitting that will equal the known degree, divide the known angle of degrees (22½) by 360⁰.

Fraction	=	Decimal
1/16	=	.0625

Q 20-18 A type-M copper pipe with a diameter of 8 inches would have an outside circumference of most nearly _____ inches.

(A) 20.12   (C) 24.28
(B) 23.13   (D) 25.13

Answer: D

Solution: 8″ x 3.1416 = 25.1328

Q 20-19 The circumference of a piece of pipe measures 18.85 inches. The diameter of the pipe would be most nearly _____ inches.

(A) 4   (C) 6
(B) 5   (D) 8

Answer: C

Solution: $\dfrac{18.85''}{3.1416}$ = 6″

Note: When the circumference is known, divide the circumference by 3.1416 to find diameter.

Q 20-20 The area of a 12-inch pipe would be most nearly _____ square inches.

(A) 113.10   (C) 114.18
(B) 110.25   (D) 111.72

Answer: A

Solution: 12″ x 12″ x .7854 = 113.0976 sq. in.

Note: To find the area of a circle (pipe), square the diameter and multiply by 0.7854 (half of π).

Q 20-21 You are installing cast iron piping and need a fitting which makes an angle of 60 degrees with the horizontal. You would select a _____ bend from the following fittings.

(A) 1/8   (C) 1/16
(B) 1/5   (D) 1/6

Answer: D

Solution: $\dfrac{60^0}{360^0}$ = 1/6-bend

Note: To find a fitting that will equal the known degree, divide the known angle of degrees (60°) by 360°. If this is done in decimal form, in this particular case, the solution would be 0.166-2/3, or 0.167, which is equivalent to 1/6 in fraction form.

20-22 Water collected on a roof 40 feet long by 25 feet wide by 4 inches deep will weigh most nearly _____ pounds.

(A) 18,562   (C) 21,000
(B) 20,812   (D) 22,448

Answer: B

Solution: 40' x 25' x 1/3' x 62.5 lb./cu. ft. = 20,812.5 lb.

20-23 Calculate the invert elevation at Point B shown in Figure 20-1, if slope of pipe is ¼'' fall per foot.

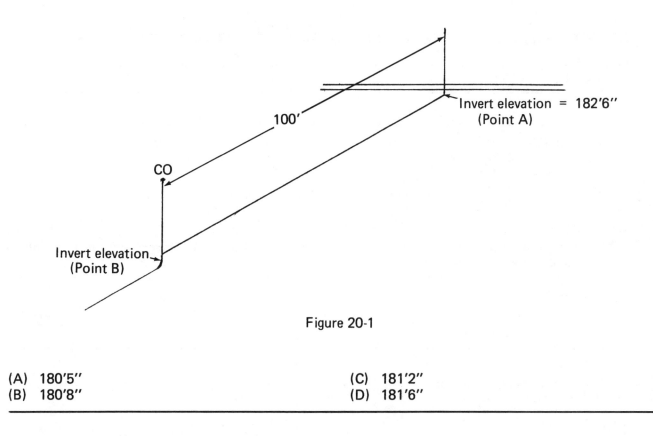

Figure 20-1

(A) 180'5''   (C) 181'2''
(B) 180'8''   (D) 181'6''

Answer: A

Solution: .25 (¼'' drop) x 100' (run) = 25'' drop (2'1'' drop)

        182'6'' − 2'1'' = 180'5'' invert elevation
        (Point A) (drop)   (Point B)

Note: The invert elevation at Point A is 182'6''. From this, subtract the drop of 25'' (2'1'').

20-24 A one-half acre parking lot with 1-inch deep standing water would hold approximately _____ gallons. Select the closest answer.

(A) 13,260

(B) 13,571

(C) 13,738

(D) 13,962

Answer: B

Solution: 21,780 sq. ft. x .0833 x 7.48 gal./cu. ft. = 13,570.769 gallons.

Note: 1 acre = 43,560 sq. ft.

½ acre = 21,780 sq. ft.

Decimal equivalent for 1/12' is 0.0833'

20-25 The amount of earth to be removed from a trench 140' x 2'6" x 4'6" is _____ cubic yards.

(A) 55.4

(B) 55.8

(C) 57.5

(D) 58.3

Answer: D

Solution: $\dfrac{140' \times 2.5' \times 4.5'}{27 \text{ cu.ft./cu.yd}} = \dfrac{1575 \text{ cu. ft.}}{27} = 58.3 \text{ cu. yd.}$

20-26 Given: A 1,200 gallon capacity grease interceptor is required by Code for a particular size restaurant. The inside width is 4'0", and the liquid depth is 4'0". The minimum inside length of the grease interceptor is _____. Select the closest answer.

(A) 9'0"

(B) 9'8"

(C) 10'0"

(D) 10'8"

Answer: C

Solution: $\dfrac{1{,}200 \text{ gal. (volume)}}{7.48 \text{ gal./cu. ft.}} = 160.43 \text{ cu. ft.}$

Area given is 4' x 4' = 16 sq. ft.

Length required is $\dfrac{160.43 \text{ cu. ft.}}{16 \text{ sq. ft.}} = 10.03 \text{ ft.}$

Size of tank is 4' x 4' x 10'

Length = 10'

Q 20-27 Assume that the full capacity of a cylindrical tank is 30 cubic feet. The tank is filled to one-half capacity with spring water. The weight of the water within the tank would be _____ pounds. Select the closest answer.

(A) 625.5

(B) 710.8

(C) 815.4

(D) 937.5

Answer: D

Solution: 15 cu. ft. (half of tank capacity) x 62.5 lb./cu.ft. = 937.5 lb.

Q 20-28 When installing PVC drainage piping requiring a 1/5-bend, the angle of turn of the fitting is _____ degrees.

(A) 40

(B) 56

(C) 60

(D) 72

Answer: D

Solution: .20 is decimal equivalent of 1/5-bend
$$.20 \times 360° = 72°$$

Note: To find the degree of angle when bend is a known factor, multiply bend x 360°.

Q 20-29 According to the data table for standard pipe threads, a 4-inch black steel pipe, threaded, would have _____ threads per inch.

(A) 8

(B) 11½

(C) 14

(D) 16

Answer: A

Solution: The standard pipe threads for pipes sized 2½" through 12" is 8 threads per inch. Refer to your math or reference book data table for standard pipe threads per inch for other pipe sizes.

Q 20-30 Assume you need to determine the offset distance for a pipe installed at an angle of 22½° with the horizontal. The run is 6 feet. Select the closest answer.

(A) 26-7/8"

(B) 27-1/8"

(C) 29-7/8"

(D) 31-9/16"

Answer: C

Solution: offset   =   .414 x 72" (6 ft.) = 29.8"

Note: When the run is known (6'), multiply run (in inches) by 0.414 to determine offset. Refer to constant table for 22½° fittings in your math or reference book.

Q 20-31 You are directed to cut a triangular opening in a wall. The opening forms a 30-60 degree right triangle. The longest side measures 10'0". The shortest side will measure most nearly ____ feet.

(A) 4

(B) 5

(C) 8

(D) 10

Answer: B

Solution: 0.5 x 10' = 5'

Note: When the diagonal (longest) side is known (10'), multiply this by 0.5 to determine run, or shortest side.

Q 20-32 Assuming the piping is standard weight, the capacity of a nominal 3-inch diameter water service pipe flowing at 8-feet per second is:

(A) between 150.0 and 175.00 gpm

(B) between 175.1 and 200.0 gpm

(C) between 200.1 and 225.0 gpm

(D) between 225.1 and 250.0 gpm

Answer: B

Solution: Multiply 0.051 sq. ft. x 8 ft./sec. x 60 sec./min. x 7.48 gal./cu. ft. = 183 gpm

Note: 0.051 is the inside cross sectional area (sq. ft.) of a 3" standard weight pipe. Refer to standard weight pipe "Diameters and Capacities" in your math or reference book.

Q 20-33 If you have pressure of 30 psi at street level, how high will this pressure lift water in a 2-inch vertical water pipe? Select the closest answer.

(A) 57.6'

(B) 68.2'

(C) 69.12'

(D) 73.10'

Answer: C

Solution: 2.304' x 30 psi = 69.12' lift

Note: Each pound per square inch of pressure requires a head of 2.304 feet. Refer to data for water pressure, head and force in your math or reference book.

Q 20-34 135 cubic feet of water weighs ____ pounds. Select the closest answer.

(A) 7,580.0

(B) 7,980.4

(C) 8,437.5

(D) 8,628.2

Answer: C

Solution: 62.5 lb. x 135 cu. ft. = 8,437.5 lb.

20-35 The foreman sends you five 10-foot and four 5-foot sections of cast iron pipe. You must install it in a system requiring 68'7" of single hub pipe. How much waste pipe will be left over?

(A) 1'5"

(B) 1'11"

(C) 2'2"

(D) 2'6"

Answer: A

Solution: 5 pieces of 10' = 50'

4 pieces of 5' = 20'

70'

Subtract 68'7" from 69'12" to get 1'5"

This 70' must be changed into 69'12" in order to subtract 68'7" from it.

Part Five
Sample Examination

- Examination Day
- First Day Exam (Morning Portion)
- First Day Exam (Afternoon Portion)
- Second Day Exam
- Answers

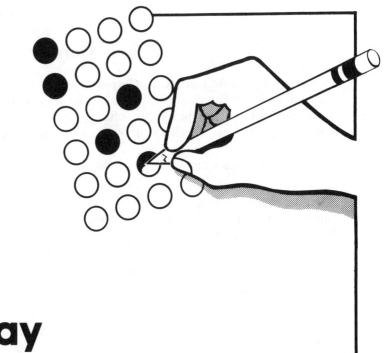

Chapter 21

Examination Day

This is your final exam. It's an open book exam, of course. Take it seriously. If you do well on this exam, you should have no trouble passing the real thing.

Don't waste this exam. This is the only sample plumber's exam you'll take before examination day. Don't try to take this sample exam before you've thoroughly studied the drawings and questions in the previous 20 chapters.

The drawings, questions and time allowed for each part of this sample test are about the same as you'll find on the real exam. The questions here are based on questions in the previous 20 chapters of this book.

Correct answers and a reference for each question are at the end of this chapter. This is for your help when *grading* the exam only.

You'll get the greatest benefit from this examination if you observe the following instructions:

• Take the practice exam on two consecutive days. Simulate test conditions. Permit no interruptions. This gives you a clearer picture of how well you can perform under time limitations.

• Follow *exactly* the instructions given at the beginning of each portion of the test.

• Either photocopy the computer answer sheet from this book or write your answers on a separate sheet of paper.

• If you can't answer a particular question, don't stop taking the exam and look up the answer. Go on to the next question. Leave correcting the exam until you've finished. Then, look up the answer in your local plumbing or gas code. Use NFPA 14, NFPA 56F and your plumbing math book for questions on medical gas and math problems. This will improve your speed in locating answers in approved references when taking the actual exam.

• When you've completed the exam, have someone else grade your exam papers, if possible. Remember, 70% is the lowest passing grade. This means you need 140 correct answers out of a total of 200 questions to pass.

• When your exam has been corrected, go back over those subjects where you had the most trouble.

If you score 80% or more on this test, you'll have little or no trouble passing the real plumber's exam. Good luck.

Plumbing Examination
First Day
(Morning Portion)

Important Instructions — Read Carefully

You will be given four hours to complete the morning portion of the examination. It consists of questions relating to general plumbing installation in a four-story building. On the following pages you will find the set of plans on which these questions are based.

The floor plans are shown as follows:

Sheet 1   Basement and first floor plan
Sheet 2   Second floor plan
Sheet 3   Typical third and fourth floor plan
Sheet 4   Roof plan
Sheet 5   Isometric for basement, first floor and kitchen

Drawings are not to scale. These drawings are for examination purposes only. A complete set of drawings and specifications is not provided. There is no intent to comply with every code. *The questions are to be answered from the plans.*

This First Day Exam (Morning Portion) will constitute 30% of your final grade. The remaining 70% will be accounted for as follows:

30% — First Day Afternoon Examination
40% — Second Day Examination

All questions on this examination carry equal weight. The minimum final passing grade is 70%.

Before the Test

Fill in all the information called for on the answer sheet. *Print* your name carefully. It is imperative that you then write your *signature* in the appropriate space provided on the answer sheet *and* at the top of this copy of the examination.

Test Instructions

Each question on this examination has four alternative answers from which you are to select *one.*

Use the separate answer sheet provided (immediately following these instructions) to record your answers. On each test item you are to fill in the circle containing the letter of the response that you choose as the correct answer. Use a Number 2 or softer pencil and blacken the circle completely. For example, if you choose "C" as the correct answer to a question, you would indicate it on the answer sheet in this manner:

You must mark only *one* answer for each question. *You will be graded only on the answers recorded on the answer sheet.*

If you make a mistake in marking your answer sheet, erase *completely* the answer you wish to remove and then mark the correct circle.

Make no stray marks on the answer sheet.

Work carefully, but do not spend too much time on any one question. It is usually better to first answer all those questions that you feel sure about and can answer quickly. Then, return to the questions you need to think about.

If you have any questions, raise your hand.

Wait for instructions to begin the test.

Plumber's Examination Answer Sheet

Name _____
Please print   (last)   (first)   (middle)

Address _____

_____

Location of Examination_____

Signature_____

Directions For Marking Answer Sheet
• Use a black lead pencil only, #2 or softer. • DO NOT use ink or ballpoint pen.
• Make heavy black marks that fill the circle completely. • ERASE cleanly any answer you wish to change.
• Make NO stray marks on this answer sheet.

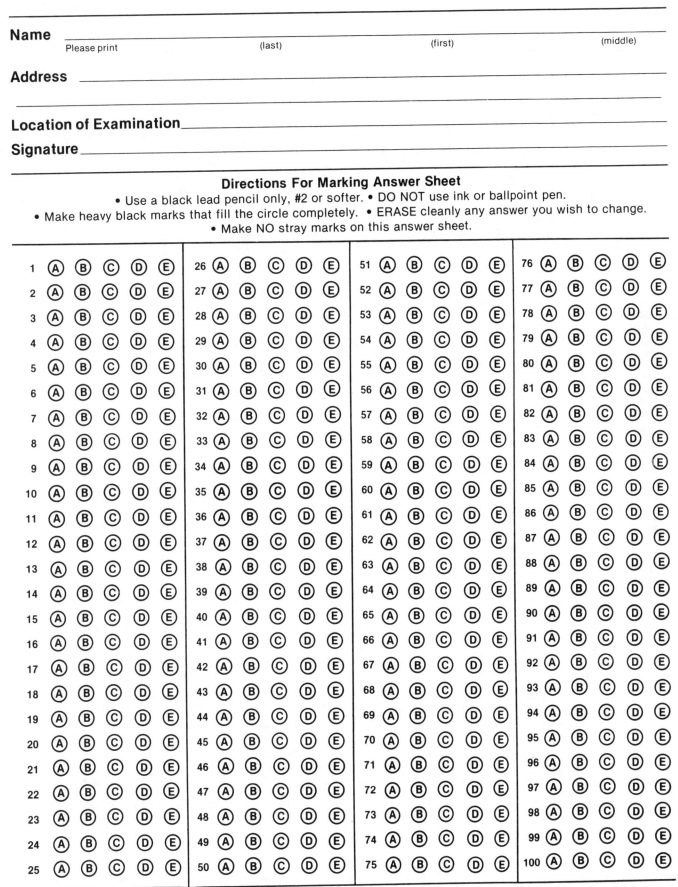

First Day (Morning Portion)

Key To Drawings

Abbreviations

BD	building drain
BS	building sewer
AW	acid waste
BSD	building storm drain
CI	cast iron
Clg	ceiling
CO	cleanout
CW	cold water
DW	dishwasher
EWC	electric water cooler
EWH	electric water heater
FC	fire cabinet
FCO	floor cleanout
FD	floor drain
FDC	fire department connection
FFE	finish floor elevatior
FL	fire line
FS	floor sink
FSP	fire standpipe
GD	garbage disposal
GWH	gas water heater
LAV	lavatory
RD	roof drain
RL	roof leader
SH	shower
SS	service sink
S	soil stack
TP	trap primer
UR	urinal
V	vent
VTR	vent through roof
WC	water closet
WCO	wall cleanout

Legend

	Fixture	Description
P-1	Water closet	floor mounted, flushometer
P-2	Handicap water closet	floor mounted, flushometer
P-3	Urinal	wall mounted, flushometer
P-4	Lavatory	counter top mounted
P-5	Service sink	2" P-trap
P-6	Electric water cooler	wall mounted
P-10	Lavatory	wall mounted
P-11	Water closet	floor mounted, tank type
P-12	Shower	tile stall
P-13	Water heater	30 gallon, 4500 watt
P-14	Hand sink	wall mounted
P-15	Glass sink	with drain board
P-16	Garbage grinder	
P-17	Coffee urn	
P-18	Floor sink	
P-20	Sink	single compartment
P-21	Commercial dishwasher	undercounter type
P-22	Pot sink	three compartment
P-23	Barber sink	wall mounted
P-24	Ice machine	
P-25	Water heater	82 gallon, natural gas
P-29	Floor drain	3" with basket
P-30	Garbage can wash	3" drain with basket
P-31	Grease interceptor	1,200 gallon capacity
P-32	Water meter	2" water meter

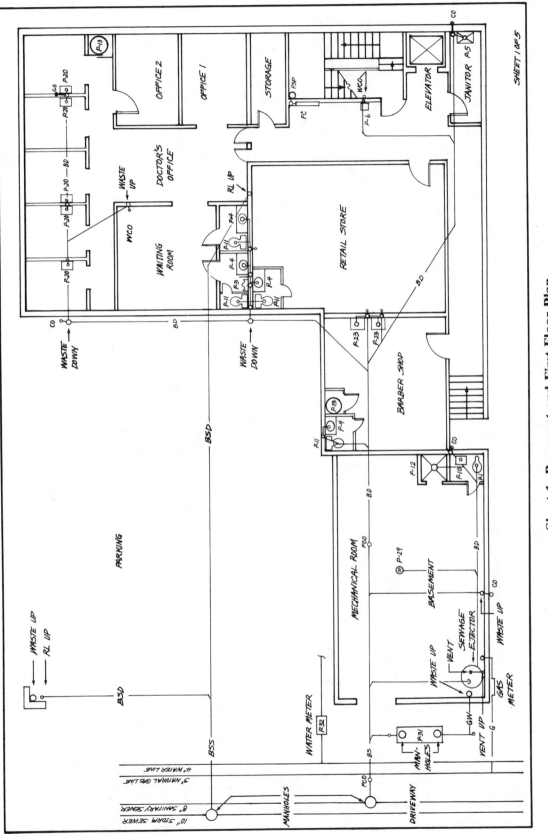

Sheet 1: Basement and First Floor Plan

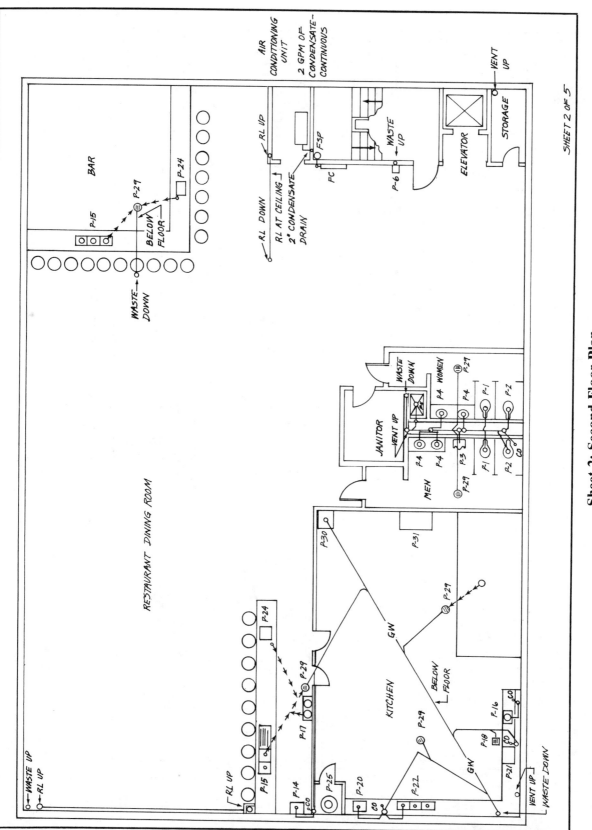

Sheet 2: Second Floor Plan

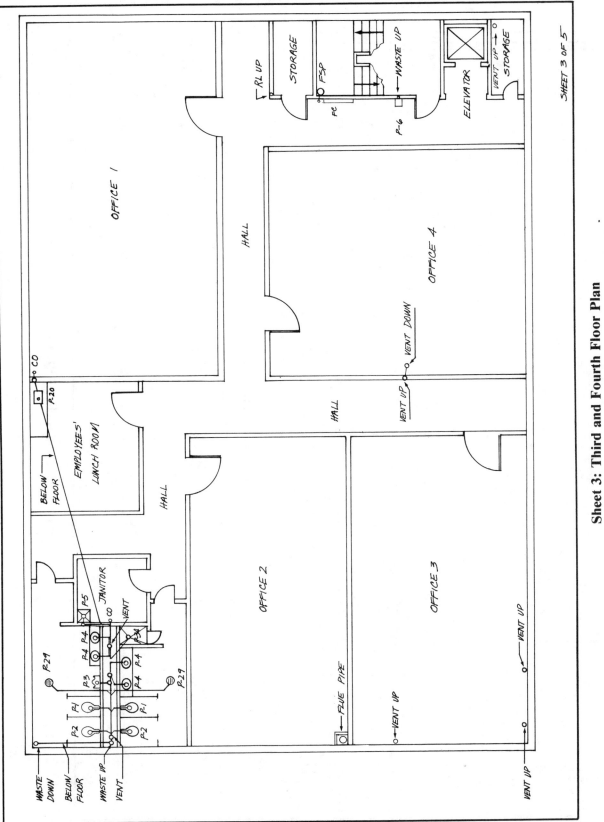

Sheet 3: Third and Fourth Floor Plan

273

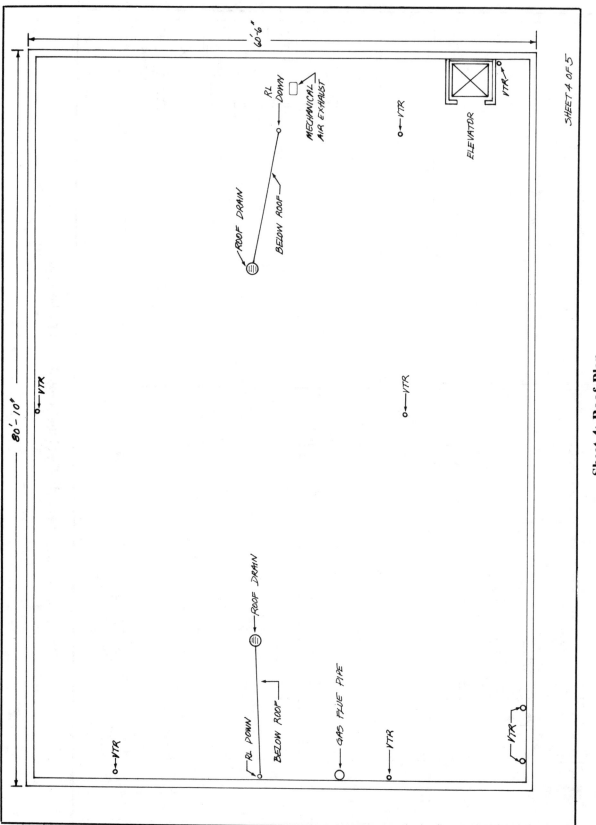

Sheet 4: Roof Plan

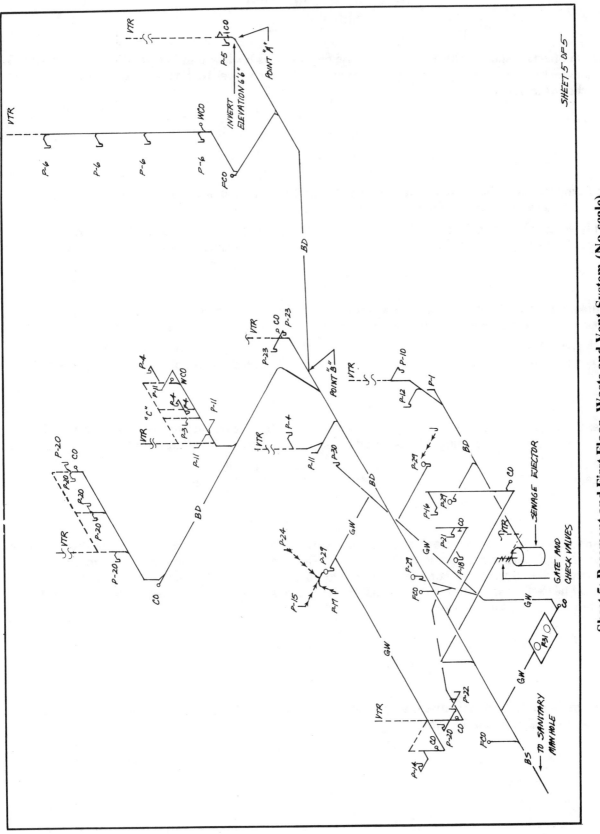

Sheet 5: Basement and First Floor, Waste and Vent System (No scale)

First Day Examination
Morning

1 Refer to Sheets 1 and 2. The gas piping for the restaurant is for low pressure gas (not in excess of 0.5 lb. per square inch). To air-test the system for inspection, Code requires that gas piping withstand a pressure of _____ inches of mercury.

(A) 2

(B) 4

(C) 6

(D) 8

2 According to Code, the gas meter as shown on Sheet 1 is located in a ventilated area (outside of building). A further Code requirement is that the meter be:

(A) readily accessible for reading

(B) preceded by a bypass connection

(C) preceded by a house shut-off valve

(D) firmly secured to outer building wall

3 Assume you have the contract for the plumbing and gas installation as shown on Sheets 1-5. Before you proceed with the gas piping installation for this restaurant, your first responsibility is to:

(A) check the job site

(B) prepare a piping diagram

(C) secure a gas permit

(D) prepare a material take-off for the job

4 Refer to Sheets 1 and 2. The job engineer's specification for the restaurant's gas piping system is nonferrous pipe with soldered joints. Code-acceptable solder for gas piping ("hard solder") is composed of:

(A) tin and lead

(B) zinc and silver

(C) copper and zinc

(D) tin, lead and zinc

5 Refer to Sheet 2. The gas water heater in this restaurant is installed in a separate room, thus requiring ventilation through permanent openings. According to Code, one vent opening should be a minimum of 12 inches above the floor. The second opening should be a minimum of 12 inches below the ceiling. If the gas water heater has an input rating of 250,000 Btu per hour, the total free area of the permanent openings must be _____ square inches.

(A) 100

(B) 250

(C) 500

(D) 550

6 Refer to Sheet 5. Code specifies that gas vent pipe must extend at least _____ inches above the highest point where it passes through the roof.

(A) 18

(B) 22

(C) 24

(D) 32

7 According to Code, the hangers used to support the vertical fire standpipe on Sheets 1-4 must be able to uphold a water-filled pipe, as well as an additional _____ pounds.

(A) 150
(B) 250

(C) 350
(D) 450

8 Code specifies that hose connections for fire hose cabinets (Sheets 1-4) be located no more than _____ feet from the finished floor.

(A) 4
(B) 5

(C) 6
(D) 7

9 The building sewer shown on Sheet 1 is 6 inches in diameter. Assume the distance is 100 feet. According to Code, the minimum drop should be most nearly _____ inches.

(A) 8.75
(B) 12.50

(C) 15.25
(D) 17.15

10 You are installing the underground sanitary systems for Sheet 1. You need a fitting which makes a 60-degree angle with the horizontal. You send your helper to the fitting pile to bring back a _____ bend.

(A) 1/8
(B) 1/5

(C) 1/16
(D) 1/6

11 Refer to Sheet 1. You will have to remove _____ cubic yards of earth from the trenches (which measure 140' x 2'6'' x 4'6'') in order to install the underground piping.

(A) 55.4
(B) 55.8

(C) 57.5
(D) 58.3

12 Refer to Sheet 2. The Code-required hot water temperature range for the dishwashing machine in the restaurant is:

(A) 110° − 120° F
(B) 130° − 140° F

(C) 170° − 180° F
(D) 190° − 200° F

13 You are designing the water system for the four-story building (Sheets 1-4). The water pressure fluctuates between a low of 40 psi in late afternoon to a high of 55 psi in mid-morning. The Code mandates that the water piping system be designed for a minimum water pressure of _____ psi.

(A) 40
(B) 45

(C) 50
(D) 55

14   Refer to Sheets 1-4. When roughing-in the waste outlet for the standard water closets (not handicapped), the center of the waste outlet must be a minimum of _____ inches from any side wall or partition.

(A)   12   (C)   18
(B)   15   (D)   24

15   The building as shown on Sheets 1-4 will employ approximately 45 workmen at peak construction. To meet Code requirements, at least _____ temporary toilet facilities (chemical toilets) must be provided for the workers' use.

(A)   1   (C)   3
(B)   2   (D)   4

16   Refer to Sheets 1-4. The Code makes it clear that all materials used in the installation of this plumbing and drainage system must conform to:

(A)   manufacturer's standards   (C)   approved maximum standards
(B)   all alternate materials available   (D)   approved minimum standards

17   The type of threads required by Code for the galvanized steel pipe used in the fire line system (Sheets 1-4) are:

(A)   Briggs standard pipe threads   (C)   WWP-401 regular pipe threads
(B)   standard taper pipe threads   (D)   WWT-791 standard pipe threads

18   The invert elevation at Point A, Sheet 5, is 6'6''. The slope of the pipe is ¼'' fall per foot and the developed length of run is 60 feet. Calculate the invert elevation at Point B.

(A)   5'3''   (C)   5'6''
(B)   5'5''   (D)   5'8''

19   The vent system C on Sheet 5 illustrates a method that may be used for venting several fixtures. The Code defines it as a:

(A)   revent   (C)   circuit vent
(B)   unit vent   (D)   loop vent

20   The roof leaders (Sheet 4) have become plugged, preventing drainage. There is a parapet wall 6 inches high. With clogged drains and continuing rains, the roof area will fill up with _____ gallons of rain water. Select the closest answer.

(A)   17,940   (C)   18,330
(B)   18,200   (D)   18,470

Plumbing Examination
First Day
(Afternoon Portion)

Important Instructions — Read Carefully

You will be given four hours to complete the afternoon portion of the examination.

Questions 1-60 of the examination are related to the interpretation of the Code, along with some math problems.

Questions 61-70 refer to the "zoned" drawing as related to the interpretation of the Code.

Please note: Regarding questions 1-70: On each test item you will have a choice of A, B, C or D.

Regarding Questions 71-80 (mini drawings): On each test item you will have a choice of *A, correct* (indicating the drawing to be *correct)* or *B, incorrect* (indicating the drawing to be *incorrect).*

This First Day Exam (Afternoon Portion) will constitute 30% of your final grade. The remaining 70% will be accounted for as follows:

30% - First Day Morning Examination
40% - Second Day Examination

All questions on this examination carry equal weight. The minimum final passing grade is 70%.

If you do not know the answer to a question, you should go ahead and mark on your answer sheet the answer you *think* is correct.

Before the Test

Fill in all the information called for on the answer sheet. *Print* your name carefully. It is imperative that you then write your *signature* in the appropriate space provided on the answer sheet *and* at the top of this copy of the examination.

Test Instructions

Each question on this examination has four alternative answers, from which you are to select *one.*

Use the separate answer sheet provided (immediately following these instructions) to record your answers. On each test item you are to fill in the circle containing the letter of the response that you choose as the correct answer. Use a Number 2 or softer pencil and blacken the circle completely. For example, if you choose "C" as the correct answer to a question, you would indicate it on the answer sheet in this manner:

You must mark only *one* answer for each question. *You will be graded only on the answers recorded on the answer sheet.*

If you make a mistake in marking your answer sheet, erase *completely* the answer you wish to remove and then mark the correct circle.

Make no stray marks on the answer sheet.

Work carefully, but do not spend too much time on any one question. It is usually better to first answer all those questions that you feel sure about and can answer quickly. Then, return to the questions you need to think about.

If you have any questions, raise your hand.

Wait for instructions to begin the test.

Plumber's Examination Answer Sheet

Name _____

Please print   (last)   (first)   (middle)

Address _____

_____

Location of Examination_____

Signature _____

Directions For Marking Answer Sheet

• Use a black lead pencil only, #2 or softer. • DO NOT use ink or ballpoint pen.
• Make heavy black marks that fill the circle completely. • ERASE cleanly any answer you wish to change.
• Make NO stray marks on this answer sheet.

1 Ⓐ Ⓑ Ⓒ Ⓓ Ⓔ	26 Ⓐ Ⓑ Ⓒ Ⓓ Ⓔ	51 Ⓐ Ⓑ Ⓒ Ⓓ Ⓔ	76 Ⓐ Ⓑ Ⓒ Ⓓ Ⓔ
2 Ⓐ Ⓑ Ⓒ Ⓓ Ⓔ	27 Ⓐ Ⓑ Ⓒ Ⓓ Ⓔ	52 Ⓐ Ⓑ Ⓒ Ⓓ Ⓔ	77 Ⓐ Ⓑ Ⓒ Ⓓ Ⓔ
3 Ⓐ Ⓑ Ⓒ Ⓓ Ⓔ	28 Ⓐ Ⓑ Ⓒ Ⓓ Ⓔ	53 Ⓐ Ⓑ Ⓒ Ⓓ Ⓔ	78 Ⓐ Ⓑ Ⓒ Ⓓ Ⓔ
4 Ⓐ Ⓑ Ⓒ Ⓓ Ⓔ	29 Ⓐ Ⓑ Ⓒ Ⓓ Ⓔ	54 Ⓐ Ⓑ Ⓒ Ⓓ Ⓔ	79 Ⓐ Ⓑ Ⓒ Ⓓ Ⓔ
5 Ⓐ Ⓑ Ⓒ Ⓓ Ⓔ	30 Ⓐ Ⓑ Ⓒ Ⓓ Ⓔ	55 Ⓐ Ⓑ Ⓒ Ⓓ Ⓔ	80 Ⓐ Ⓑ Ⓒ Ⓓ Ⓔ
6 Ⓐ Ⓑ Ⓒ Ⓓ Ⓔ	31 Ⓐ Ⓑ Ⓒ Ⓓ Ⓔ	56 Ⓐ Ⓑ Ⓒ Ⓓ Ⓔ	81 Ⓐ Ⓑ Ⓒ Ⓓ Ⓔ
7 Ⓐ Ⓑ Ⓒ Ⓓ Ⓔ	32 Ⓐ Ⓑ Ⓒ Ⓓ Ⓔ	57 Ⓐ Ⓑ Ⓒ Ⓓ Ⓔ	82 Ⓐ Ⓑ Ⓒ Ⓓ Ⓔ
8 Ⓐ Ⓑ Ⓒ Ⓓ Ⓔ	33 Ⓐ Ⓑ Ⓒ Ⓓ Ⓔ	58 Ⓐ Ⓑ Ⓒ Ⓓ Ⓔ	83 Ⓐ Ⓑ Ⓒ Ⓓ Ⓔ
9 Ⓐ Ⓑ Ⓒ Ⓓ Ⓔ	34 Ⓐ Ⓑ Ⓒ Ⓓ Ⓔ	59 Ⓐ Ⓑ Ⓒ Ⓓ Ⓔ	84 Ⓐ Ⓑ Ⓒ Ⓓ Ⓔ
10 Ⓐ Ⓑ Ⓒ Ⓓ Ⓔ	35 Ⓐ Ⓑ Ⓒ Ⓓ Ⓔ	60 Ⓐ Ⓑ Ⓒ Ⓓ Ⓔ	85 Ⓐ Ⓑ Ⓒ Ⓓ Ⓔ
11 Ⓐ Ⓑ Ⓒ Ⓓ Ⓔ	36 Ⓐ Ⓑ Ⓒ Ⓓ Ⓔ	61 Ⓐ Ⓑ Ⓒ Ⓓ Ⓔ	86 Ⓐ Ⓑ Ⓒ Ⓓ Ⓔ
12 Ⓐ Ⓑ Ⓒ Ⓓ Ⓔ	37 Ⓐ Ⓑ Ⓒ Ⓓ Ⓔ	62 Ⓐ Ⓑ Ⓒ Ⓓ Ⓔ	87 Ⓐ Ⓑ Ⓒ Ⓓ Ⓔ
13 Ⓐ Ⓑ Ⓒ Ⓓ Ⓔ	38 Ⓐ Ⓑ Ⓒ Ⓓ Ⓔ	63 Ⓐ Ⓑ Ⓒ Ⓓ Ⓔ	88 Ⓐ Ⓑ Ⓒ Ⓓ Ⓔ
14 Ⓐ Ⓑ Ⓒ Ⓓ Ⓔ	39 Ⓐ Ⓑ Ⓒ Ⓓ Ⓔ	64 Ⓐ Ⓑ Ⓒ Ⓓ Ⓔ	89 Ⓐ Ⓑ Ⓒ Ⓓ Ⓔ
15 Ⓐ Ⓑ Ⓒ Ⓓ Ⓔ	40 Ⓐ Ⓑ Ⓒ Ⓓ Ⓔ	65 Ⓐ Ⓑ Ⓒ Ⓓ Ⓔ	90 Ⓐ Ⓑ Ⓒ Ⓓ Ⓔ
16 Ⓐ Ⓑ Ⓒ Ⓓ Ⓔ	41 Ⓐ Ⓑ Ⓒ Ⓓ Ⓔ	66 Ⓐ Ⓑ Ⓒ Ⓓ Ⓔ	91 Ⓐ Ⓑ Ⓒ Ⓓ Ⓔ
17 Ⓐ Ⓑ Ⓒ Ⓓ Ⓔ	42 Ⓐ Ⓑ Ⓒ Ⓓ Ⓔ	67 Ⓐ Ⓑ Ⓒ Ⓓ Ⓔ	92 Ⓐ Ⓑ Ⓒ Ⓓ Ⓔ
18 Ⓐ Ⓑ Ⓒ Ⓓ Ⓔ	43 Ⓐ Ⓑ Ⓒ Ⓓ Ⓔ	68 Ⓐ Ⓑ Ⓒ Ⓓ Ⓔ	93 Ⓐ Ⓑ Ⓒ Ⓓ Ⓔ
19 Ⓐ Ⓑ Ⓒ Ⓓ Ⓔ	44 Ⓐ Ⓑ Ⓒ Ⓓ Ⓔ	69 Ⓐ Ⓑ Ⓒ Ⓓ Ⓔ	94 Ⓐ Ⓑ Ⓒ Ⓓ Ⓔ
20 Ⓐ Ⓑ Ⓒ Ⓓ Ⓔ	45 Ⓐ Ⓑ Ⓒ Ⓓ Ⓔ	70 Ⓐ Ⓑ Ⓒ Ⓓ Ⓔ	95 Ⓐ Ⓑ Ⓒ Ⓓ Ⓔ
21 Ⓐ Ⓑ Ⓒ Ⓓ Ⓔ	46 Ⓐ Ⓑ Ⓒ Ⓓ Ⓔ	71 Ⓐ Ⓑ Ⓒ Ⓓ Ⓔ	96 Ⓐ Ⓑ Ⓒ Ⓓ Ⓔ
22 Ⓐ Ⓑ Ⓒ Ⓓ Ⓔ	47 Ⓐ Ⓑ Ⓒ Ⓓ Ⓔ	72 Ⓐ Ⓑ Ⓒ Ⓓ Ⓔ	97 Ⓐ Ⓑ Ⓒ Ⓓ Ⓔ
23 Ⓐ Ⓑ Ⓒ Ⓓ Ⓔ	48 Ⓐ Ⓑ Ⓒ Ⓓ Ⓔ	73 Ⓐ Ⓑ Ⓒ Ⓓ Ⓔ	98 Ⓐ Ⓑ Ⓒ Ⓓ Ⓔ
24 Ⓐ Ⓑ Ⓒ Ⓓ Ⓔ	49 Ⓐ Ⓑ Ⓒ Ⓓ Ⓔ	74 Ⓐ Ⓑ Ⓒ Ⓓ Ⓔ	99 Ⓐ Ⓑ Ⓒ Ⓓ Ⓔ
25 Ⓐ Ⓑ Ⓒ Ⓓ Ⓔ	50 Ⓐ Ⓑ Ⓒ Ⓓ Ⓔ	75 Ⓐ Ⓑ Ⓒ Ⓓ Ⓔ	100 Ⓐ Ⓑ Ⓒ Ⓓ Ⓔ

First Day (Afternoon Portion)

First Day Examination
Afternoon

1 According to the Code, it shall be the duty and responsibility of the _____ to determine if the plumbing has been inspected before it is covered or concealed.

(A) general contractor

(B) permit holder

(C) plumbing inspector

(D) building and zoning record control branch

2 The Code requires that when a water service pipe is installed in the same trench with a building sewer, there must be a minimum separation of _____ inches.

(A) 6

(B) 10

(C) 12

(D) 18

3 In existing buildings in which plumbing installations are to be renovated, the Code may permit the work to be done:

(A) only in a workmanlike manner

(B) with necessary deviations, provided the intent of the Code is met

(C) if like materials are used for replacement

(D) without a permit

4 In lieu of the water test, the Code will accept one of the following for a final test of a completed drainage and vent system:

(A) air

(B) gas mixture of oil of peppermint

(C) oxygen

(D) mercury vapor

5 Whenever compliance with all the provisions of the Code fails to eliminate _____, the owner or his agent have to make acceptable corrections.

(A) bad workmanship

(B) a drainage problem

(C) a nuisance

(D) uncertified workers from a job

6 The Code specifies that all trenching required for the installation of a plumbing system within the walls of a building be:

(A) not closer than 5 feet to a building foundation

(B) not deeper than 3 feet maximum

(C) covered immediately after installation for safety of other workers

(D) open trench work

7 The Code states that 3-inch PVC waste piping must not be installed in outside walls of a building:

(A) unless wall is a minimum of 6 inches in width

(B) if temperatures of the area ever exceed 100 degrees F.

(C) if temperatures of the area are ever less than 32 degrees F.

(D) if the wall is constructed of combustible material

First Day Examination
Afternoon

8 According to Code, a water test may be applied to the drainage and vent system of a two-story residence:

(A) only in its entirety

(B) only in sections

(C) only during working hours

(D) in its entirety or in sections

9 A horizontal waste and vent pipe which is sized to provide free circulation of air above the flow line of the drain pipe is defined by Code as a:

(A) combination waste and vent system

(B) continuous wet vent system

(C) non-solvent waste and vent system

(D) ventilating piping system

10 A vent pipe which receives waste from a shower is defined by Code as:

(A) a wet vent

(B) a common vent

(C) a circuit vent

(D) an island vent

11 The gradient of a horizontal line of pipe, as defined by Code, is usually expressed in terms of:

(A) a fraction of an inch

(B) a fraction of a foot

(C) being below the frost line

(D) its extreme depth

12 The Code defines any horizontal pipe or fittings as those installed:

(A) with a pitch of not more than 1/8 inch fall per foot

(B) with a slope in line of a pipe, in reference to a horizontal plane

(C) with an angle of more than 45° with the vertical

(D) with an angle of less than 45° with the vertical

13 The Code requires that soldering bushings used with 2½-inch pipe have a minimum weight of _____ ounces.

(A) 6

(B) 8

(C) 14

(D) 22

14 To meet Code, floor flanges for water closets manufactured of cast iron must have a minimum thickness of _____ inch.

(A) 1/16

(B) 1/8

(C) 3/16

(D) 1/4

15 Where sheet lead is used to make watertight vent terminals, the Code requires that such sheet lead weigh no less than _____ pounds per square foot.

(A) 2

(B) 2½

(C) 3

(D) 3½

16 Polybutylene piping and fittings, according to Code, would most likely be used in:

(A) a water distribution system

(B) special wastes systems

(C) indirect waste piping

(D) wet-vented systems

17 Of the following listed materials, the one approved by Code to secure water closets to tile floors is:

(A) brass

(B) Bismuth

(C) Babbitt metal

(D) tungsten

18 The Code requirement is that 4-inch pipe size caulking ferrules weigh a minimum of _____ each.

(A) 1 lb. 6 oz.

(B) 1 lb. 12 oz.

(C) 2 lb. 3 oz.

(D) 2 lb. 8 oz.

19 By Code requirements, a trap depending on movable parts to retain its seal may:

(A) be used for fixtures with clear water waste only

(B) not be used under any circumstances

(C) be used if first approved by the plumbing official

(D) be used when approved by the ASME Standards

20 Given: Several adjacent lavatories are installed on the same wall in the same room. According to Code, _____ lavatories may use a single trap.

(A) 2

(B) 3

(C) 4

(D) 5

21 According to Code, horizontal combination waste-and-vent sanitary systems are limited to:

(A) water closets only

(B) floor drains only

(C) floor sinks only

(D) fixtures not adjacent to walls or partitions

22 Fixture unit equivalents not commonly listed in fixture unit tables, according to Code, must be based on the:

(A) type of fixture
(B) trap size

(C) location of fixture
(D) discharge capacity of fixture

23 Assume that the waste is discharged from a building at the rate of 630 gpm. The rate of flow in terms of fixture units is most nearly:

(A) 62
(B) 76

(C) 84
(D) 108

24 The flow from a pump ejector in a building is 66 gpm. According to Code, this would add a total of _____ fixture units to the building drainage system.

(A) 123
(B) 128

(C) 133
(D) 137

25 A 4-inch sewer pipe, according to Code, must have a fall no less than _____ inches every 50 feet.

(A) 5.5
(B) 6.0

(C) 6.25
(D) 6.75

26 According to Code, of the following plumbing piping, the one in which hoarfrost often collects is the:

(A) soil stack
(B) vent stack

(C) waste stack
(D) cold water service

27 One of the following materials is acceptable by Code for venting a chemical waste system. It is:

(A) type-L copper pipe
(B) PVC Schedule 40 pipe

(C) galvanized wrought iron pipe
(D) borosilicate glass

28 Interceptors, according to Code, must be designed:

(A) so they will not become air bound
(B) of concrete to prevent corrosion

(C) for easy use of ladders to service interceptors
(D) with an approved type flow control device

29 According to Code, interceptors have to be installed:

(A) by a licensed plumbing contractor
(B) by a licensed septic tank contractor

(C) where they are accessible
(D) next to a three-compartment pot sink

30 The requirements for installation of grease interceptors, according to Code, do <u>not</u> include:

(A) fast food restaurants

(B) apartment buildings with up to 100 dwelling units

(C) hospitals

(D) hotels

31 To properly connect a steam pipe to a plumbing drainage system, it is mandated by Code that:

(A) the pipe be not less than 10 feet in length

(B) the diameter of the pipe be a minimum of 1¼ inches

(C) a direct connection be made to a waste stack only

(D) an indirect connection be made

32 When designing a horizontal waste and vent system, the waste piping, according to Code, must be:

(A) at least 3 inches in diameter

(B) graded a minimum of ¼-inch fall per foot

(C) at least two pipe sizes larger than conventional pipe sizes

(D) no less than standard weight pipe

33 According to Code, in designing the liquid capacity of septic tanks for a strip store shopping center, the determining factor in sizing is the:

(A) number of persons

(B) plumbing fixture units

(C) number of toilet rooms

(D) engineering data available

34 When the Code permits the installation of a disposal field, the minimum square feet of trench bottom has to be at least _____ square feet.

(A) 100

(B) 125

(C) 150

(D) 175

35 Given: A rural single family residence is to be constructed on an extra large lot. A small stream crosses the property. The sewage disposal facilities consist of a septic tank and disposal field. The minimum horizontal distance between the septic tank and the stream, according to Code, can be no less than _____ feet.

(A) 50

(B) 75

(C) 100

(D) 150

36 The plumber, according to Code, is responsible for providing adequate protection to prevent the water system from possible:

(A) temperature variation

(B) stresses

(C) back siphonage

(D) pressure drops

37   When the residual pressure in a water main is not adequate to provide the minimum water pressure for plumbing fixtures to function properly at the highest water outlet, the Code mandates that the plumber install:

(A)   the next larger pipe size for the building water service line

(B)   a water pressure pump

(C)   an approved engineered water hammer arrester on top of most pipes

(D)   a pressure tank in the building water service line

38   To meet Code requirements in Vero Beach, Florida, hose bibbs installed on a single family residence must:

(A)   be a maximum of ½-inch pipe diameter

(B)   exit from the building wall a minimum of 12 inches above grade

(C)   be protected from freezing

(D)   be protected by an approved non-removable type backflow prevention device

39   According to Code, a roof drain strainer must have a minimum inlet area ____ time(s) the pipe to which it is connected.

(A)   1

(B)   1½

(C)   2

(D)   2½

40   Given: You are installing a rain water leader in a parking garage serving a six-story office building. It is firmly secured to a column supporting the upper floors. According to Code:

(A)   it should not be secured to the supporting column

(B)   it should be relocated to the inside wall

(C)   it must be protected

(D)   the material for the rain water leader must be no less than cast iron pipe

41   Rain water piping installed within a vent or shaft of a building, according to Code, shall <u>not</u> be:

(A)   Schedule 40 PVC plastic piping

(B)   26 gauge galvanized sheet metal piping

(C)   cast iron piping

(D)   galvanized steel piping

42   Which of the following statements is most nearly correct, according to Code?

(A)   Vertical leaders may be decreased in size before passing through the roof

(B)   Vertical leaders shall be sized for the maximum projected roof area

(C)   Rain leaders shall be sized in accordance with the type roof drains required

(D)   Under no circumstances shall rain leaders exceed the minimum pipe sizes as required by Code

43 Copper tubing is made in four weights, called types. According to Code, which of the following statements is most nearly correct?

(A) DWV copper pipe comes in approximately 20 foot lengths

(B) The minimum weight for rain leaders is no less than type-L copper

(C) Rain water leaders installed of copper pipe need not be water tested

(D) Solder used to make joints in a copper rain water piping system shall be no less than 60-40.

44 According to the Code, a commercial dishwashing machine must be:

(A) directly connected to a building sanitary drain

(B) indirectly connected to a building sanitary drain

(C) directly connected to a building greasy waste line

(D) indirectly connected to a building greasy waste line

45 You are estimating the lavatory needs for an industrial building. Circular basin floor-mounted fixtures are to be used. According to Code, each _____ inches is considered equivalent to one lavatory.

(A) 12

(B) 18

(C) 20

(D) 24

46 Sometimes workers in industrial plants are exposed to skin contamination from irritating materials. Where this type of work condition exists, the Code requires that one lavatory be installed for each _____ persons.

(A) 2

(B) 3

(C) 4

(D) 5

47 You are estimating the urinal and water closet needs for common toilet facilities in a commercial building for multiple tenant use. The architect has specified the need for 12 water closets. In lieu of the 12 water closets, you could comply with Code by installing:

(A) 1 urinal and 11 water closets

(B) 2 urinals and 10 water closets

(C) 3 urinals and 9 water closets

(D) 5 urinals and 7 water closets

48 According to Code, shower drains must:

(A) be constructed of cast iron with chrome strainers

(B) have adequate weep holes

(C) be set level with the shower floor

(D) be no smaller than 1½ inches in diameter

First Day Examination
Afternoon

49 The Code mandates that a water closet for conventional use be set no closer than 15 inches from the center of the bowl to any wall or partition. However, ANSI mandates that water closets for the physically handicapped must have a minimum distance of _____ inches.

(A) 16   (C) 18
(B) 17   (D) 20

50 Toilet rooms for men usually have wall-mounted urinals. The opening of a standard-mounted urinal from the floor is approximately 24 inches. According to ANSI, the opening of a urinal for the handicapped must be no higher than _____ inches from the floor.

(A) 19   (C) 21
(B) 20   (D) 22

51 If 600 feet of pipe weighs 900 pounds, 140 feet of like pipe would weigh _____ pounds. Select the closest answer.

(A) 180   (C) 211
(B) 195   (D) 217

52 A 60-gallon water heater weighs 52.5 pounds when empty and _____ pounds when full. Select the closest answer.

(A) 396   (C) 553
(B) 501   (D) 585

53 A 4-inch soil stack is approximately 60 feet high. If all openings were plugged and the stack filled with water, the pressure at the base of the stack would equal _____ pounds per square inch. Select the closest answer.

(A) 21.07   (C) 24.12
(B) 23.25   (D) 26.01

54 Given: The architect specifies sheet copper safe pans for all showers within a building. Inside dimensions of shower stalls are 3' x 4' with a 6" turnup. According to Code each shower safe pan would weigh _____ pounds.

(A) 12   (C) 14
(B) 13   (D) 15

First Day Examination
Afternoon

55 Given: A roof that is 100 feet long, 50 feet wide, with a parapet wall 6 inches high. If roof leaders should become clogged and fill the roof to overflowing, the weight of the rain water on the roof would be most nearly _____ pounds.

(A) 150,700

(B) 156,250

(C) 180,250

(D) 185,700

56 You are cutting four pieces of pipe end-to-end, with the following dimensions: 3'2-1/4", 4'7-3/8", 5'4-5/8", 6'6-7/8". To make these cuts, you would need _____ of pipe. Select the closest answer.

(A) 19'6"

(B) 19'10-1/8"

(C) 19'9"

(D) 20'2-3/16"

57 Offsets in water piping can be made with a number of acceptable fittings. If the constant 2.613 is used to compute the developed length of the offset piping, then the fittings used to make the offset must be:

(A) 5-5/8°

(B) 11-1/4°

(C) 22-1/2°

(D) 30°

58 The circumference of a piece of pipe measures 18.85 inches. The diameter of the pipe would be most nearly _____ inches.

(A) 4

(B) 5

(C) 6

(D) 8

59 If you had pressure of 30 psi at street level, how high will this pressure lift water in a 2" vertical water pipe? Select the closest answer.

(A) 57.6

(B) 68.2

(C) 69.12

(D) 73.10

60 Given: A 1,200-gallon capacity grease interceptor is required by Code for a particular size restaurant. The inside width is 4'0", and the liquid depth is 4'0". The minimum inside length of the grease interceptor is _____ . Select the closest answer.

(A) 9'0"

(B) 9'8"

(C) 10'0"

(D) 10'8"

Use this drawing to answer questions 61 through 70

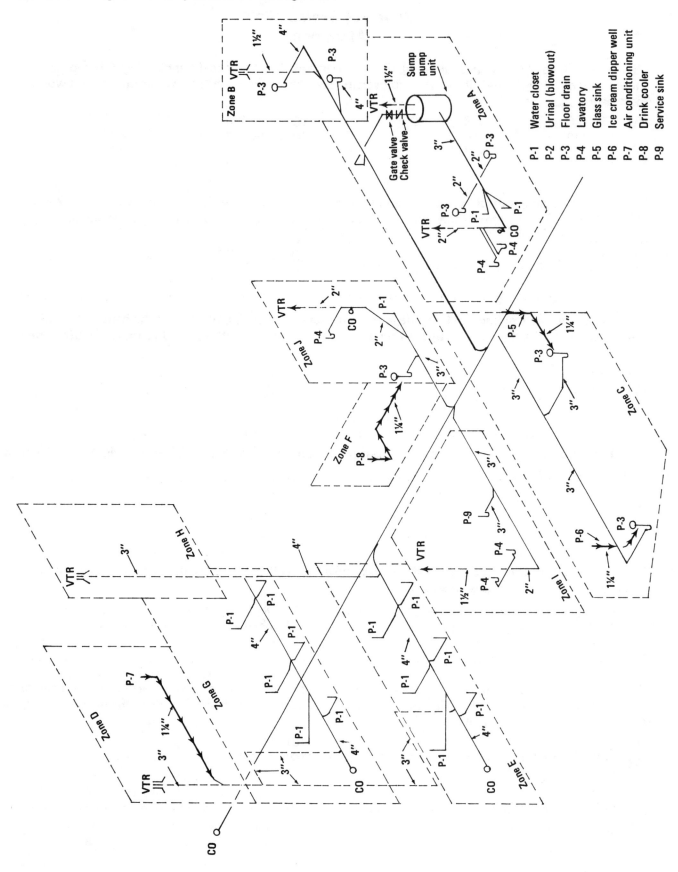

61   In the zoned drawing on the previous page, zone A shows a simple basement drainage system discharging into a sump pumping unit which connects into the drainage system. According to Code, the drawing is:

(A)   incorrect. To meet Code the sump vent should be a minimum of 3" in diameter

(B)   correct. The system as illustrated complies with the Code

(C)   incorrect. The connection to the building drain does not comply with the Code

(D)   incorrect. The gate valve and check valve installation should be reversed

62   Zone B shows two floor drains which connect into the building drainage system. According to Code, the drawing is:

(A)   incorrect. The floor drains must be a minimum of 3" in size

(B)   correct. The vent pipe location complies with Code but should have a 2" diameter

(C)   correct. The vent pipe location and size comply with Code

(D)   incorrect. The vent pipe should be located downstream from the last floor drain connection

63   Zone C reflects the roughing-in of the waste piping for an ice cream parlor. According to Code, the drawing is:

(A)   correct. The system as illustrated complies with Code

(B)   correct. The system as illustrated complies with pipe sizing requirements of the Code but should be vented

(C)   correct. However, the 3" pipe sizes can be reduced to 2" diameters and still comply with Code

(D)   incorrect. Branch venting is required for this type system

64   Zone D shows the condensate drain from an A.C. unit discharging into the main stack vent for the building. According to Code, the drawing is:

(A)   incorrect. The vent pipe must have a minimum 4" diameter to receive discharge from the A.C. unit

(B)   correct. The drawing as illustrated complies with Code

(C)   incorrect. The A.C. condensate drain should connect into the vent pipe below the loop vent connection

(D)   incorrect. The A.C. condensate drain should be indirectly connected into the drainage system

65   Zone E shows a battery of water closets. According to Code, the drawing is:

(A)   correct. The drawing as illustrated complies with Code

(B)   incorrect. A 3" vent must be installed downstream from the last water closets

(C)   correct. However, the vent pipe size can be reduced to 2" in diameter

(D)   incorrect. A relief vent is required downstream from the two floor drain connections to the horizontal drain

66 Zone F shows a floor drain receiving an indirect waste connection. According to Code, the drawing is:

(A) correct. However, the indirect waste pipe may be reduced one pipe size in diameter

(B) incorrect. To meet Code requirements, the indirect waste pipe must be vented

(C) correct. The system as illustrated complies with the pipe sizes required by Code

(D) incorrect. The system as illustrated is incorrectly connected to the drainage system

67 Zone G shows a battery of water closets. According to Code, the drawing is:

(A) correct. The system as illustrated complies with Code

(B) incorrect. Floor mounted fixtures must have a vent installed downstream from the last two fixtures

(C) incorrect. Floor mounted fixtures as illustrated must have a vent installed between the last two fixtures upstream

(D) correct. The vent piping size complies with Code

68 Zone H shows the main vent for two floors of public toilet rooms. According to Code, the drawing is:

(A) incorrect. The vent should be 4" in diameter

(B) correct. The drawing as illustrated complies with Code

(C) correct. However, the vent pipe may be one pipe size smaller and still comply with Code

(D) incorrect. The vent pipe should be offset at this point

69 Zone I shows a simple sanitary isometric. According to Code, the drawing is:

(A) incorrect. The waste pipe should be 3" up to the lavatory connection

(B) incorrect. The Code prohibits the discharge of a service sink into a wet vent system

(C) incorrect. The vent pipe should have a minimum 2" diameter

(D) correct. The drawing as illustrated complies with Code

70 Zone J illustrates the sanitary isometric of a single station toilet room. According to Code, the drawing is:

(A) incorrect. The wet vent should be a minimum of 2½" in diameter

(B) correct. However, the vent pipe can be increased one pipe size

(C) incorrect. The wet vent pipe can be reduced by one pipe size and still comply with Code

(D) correct. The drawing as illustrated complies with Code

71 According to Code the drawing as shown below is ____.

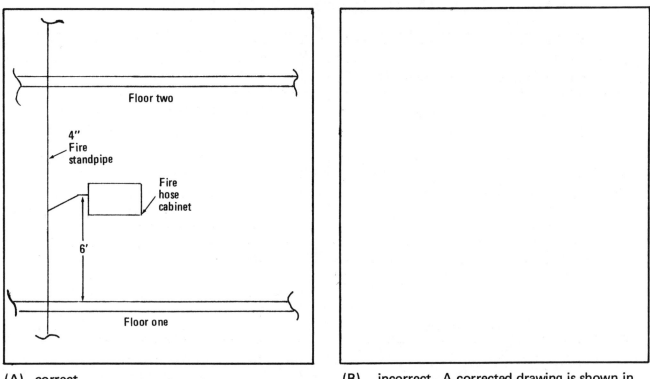

(A) correct

(B) incorrect. A corrected drawing is shown in the space provided above

72 According to Code the drawing as shown below is ____.

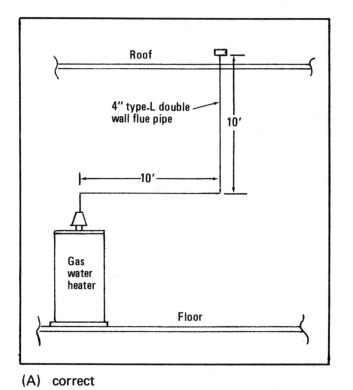

(A) correct

(B) incorrect. A corrected drawing is shown in the space provided above

73 According to Code, the drawing shown below is _____ . Pipe size is __not__ a factor.

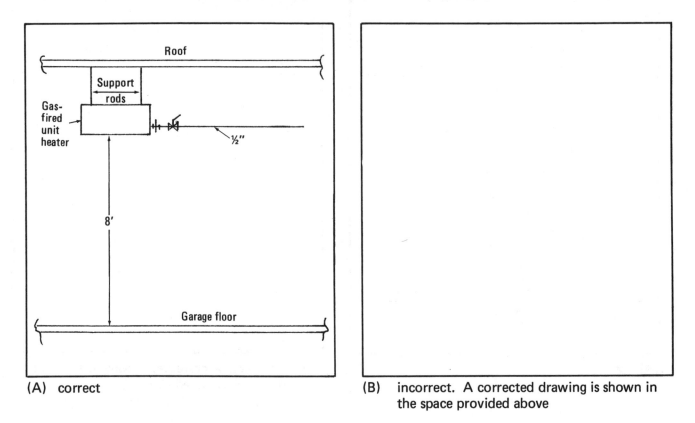

(A)   correct

(B)   incorrect. A corrected drawing is shown in the space provided above

74 According to Code, the isometric drawing shown below is _____. Pipe size is __not__ a factor.

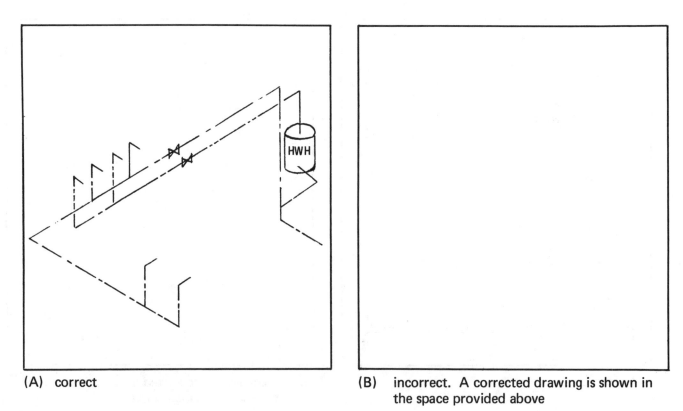

(A)   correct

(B)   incorrect. A corrected drawing is shown in the space provided above

75 According to Code the isometric drawing shown below is _____ . Pipe size <u>is</u> a factor.

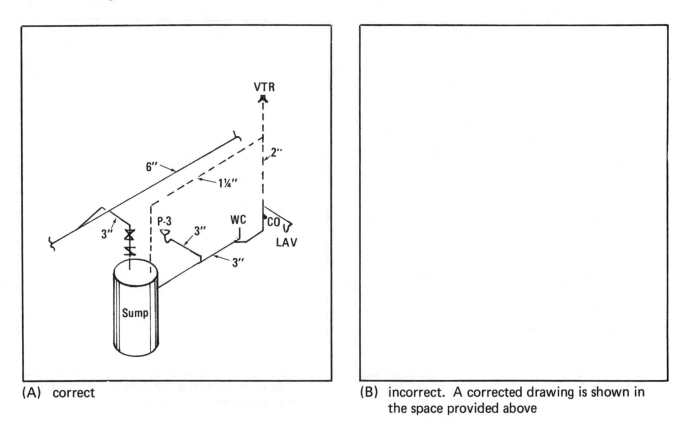

(A) correct

(B) incorrect. A corrected drawing is shown in the space provided above

76 According to Code, the isometric drawing shown below is _____ . Pipe size <u>is</u> a factor.

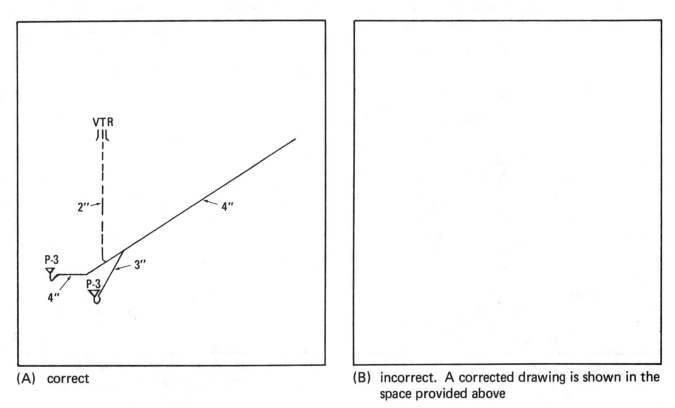

(A) correct

(B) incorrect. A corrected drawing is shown in the space provided above

77 According to Code, the piping diagram shown below is ____. Pipe size and length are _not_ factors.

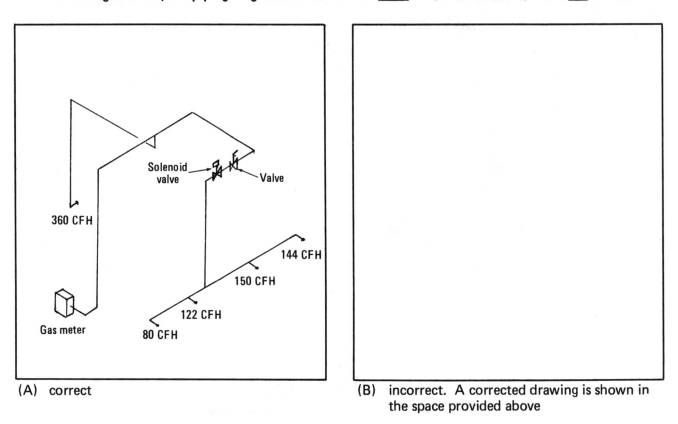

360 CFH

Solenoid valve — Valve

144 CFH

150 CFH

122 CFH

Gas meter

80 CFH

(A) correct

(B) incorrect. A corrected drawing is shown in the space provided above

78 According to Code, the isometric drawing shown below is ____. Pipe size _is_ a factor.

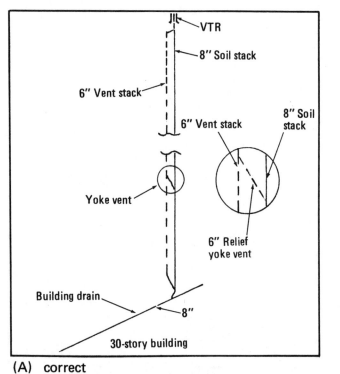

VTR

8" Soil stack

6" Vent stack

6" Vent stack

8" Soil stack

Yoke vent

6" Relief yoke vent

Building drain

8"

30-story building

(A) correct

(B) incorrect. A corrected drawing is shown in the space provided above

79 According to Code, the isometric drawing shown below is _____ . Pipe size **is** a factor.

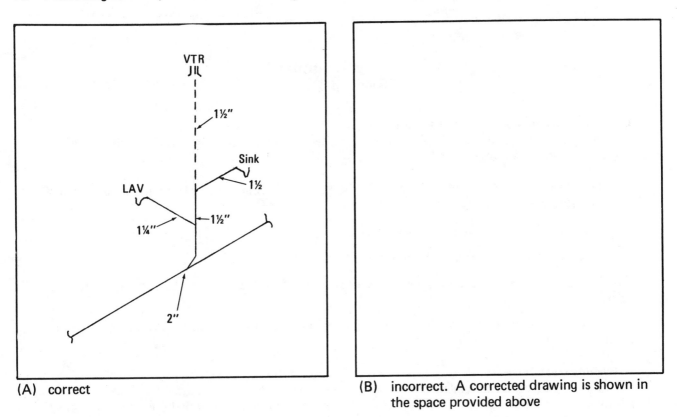

(A) correct

(B) incorrect. A corrected drawing is shown in the space provided above

80 According to Code, the isometric drawing shown below is _____ . Pipe size is **not** a factor.

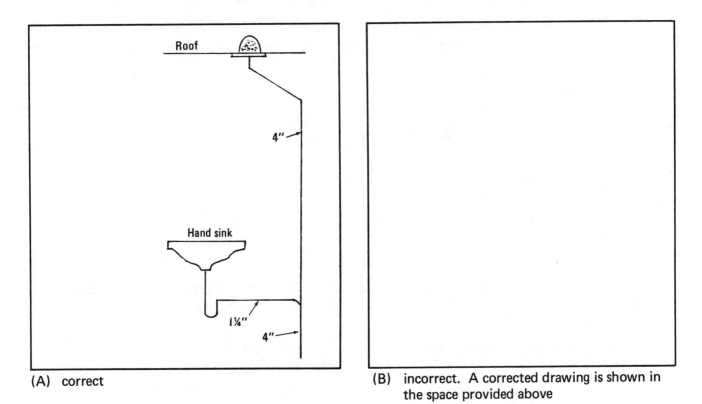

(A) correct

(B) incorrect. A corrected drawing is shown in the space provided above

Plumbing Examination
Second Day

Important Instructions — Read Carefully
Following are 100 questions related to various codes. You will have five hours to complete this test. There will be no break for lunch.

This Second Day Examination constitutes 40% of your final grade. The remaining 60% will be accounted for as follows:

30% - First Day Morning Examination

30% - First Day Afternoon Examination

All questions on the examination carry equal weight. The minimum final passing grade is 70%.

If you do not know the answer to a question, you should go ahead and mark on your answer sheet the answer you *think* is correct.

Before the Test
Fill in all the information called for on the answer sheet. *Print* your name carefully. It is imperative that you then write your *signature* in the appropriate space provided on the answer sheet *and* at the top of this copy of the examination.

Test Instructions
Each question on this examination has four alternative answers, from which you are to select *one*.

Use the separate answer sheet provided (immediately following these instructions) to record your answers. On each test item you are to fill in the circle containing the letter of the response that you choose as the correct answer. Use a Number 2 or softer pencil and blacken the circle completely. For example, if you choose "C" as the correct answer to a question, you would indicate it on the answer sheet in this manner:

You must mark only *one* answer for each question. *You will be graded only on the answers recorded on the answer sheet.*

If you make a mistake in marking your answer sheet, erase *completely* the answer you wish to remove and then mark the correct circle.

Make no stray marks on the answer sheet.

Work carefully, but do not spend too much time on any one question. It is usually better to first answer all those questions that you feel sure about and can answer quickly. Then, return to the questions you need to think about.

If you have any questions, raise your hand.

Wait for instructions to begin the test.

Plumber's Examination Answer Sheet

Name _____
  Please print   (last)   (first)   (middle)

Address _____

_____

Location of Examination_____

Signature _____

Directions For Marking Answer Sheet
• Use a black lead pencil only, #2 or softer. • DO NOT use ink or ballpoint pen.
• Make heavy black marks that fill the circle completely. • ERASE cleanly any answer you wish to change.
• Make NO stray marks on this answer sheet.

1 (A) (B) (C) (D) (E)	26 (A) (B) (C) (D) (E)	51 (A) (B) (C) (D) (E)	76 (A) (B) (C) (D) (E)
2 (A) (B) (C) (D) (E)	27 (A) (B) (C) (D) (E)	52 (A) (B) (C) (D) (E)	77 (A) (B) (C) (D) (E)
3 (A) (B) (C) (D) (E)	28 (A) (B) (C) (D) (E)	53 (A) (B) (C) (D) (E)	78 (A) (B) (C) (D) (E)
4 (A) (B) (C) (D) (E)	29 (A) (B) (C) (D) (E)	54 (A) (B) (C) (D) (E)	79 (A) (B) (C) (D) (E)
5 (A) (B) (C) (D) (E)	30 (A) (B) (C) (D) (E)	55 (A) (B) (C) (D) (E)	80 (A) (B) (C) (D) (E)
6 (A) (B) (C) (D) (E)	31 (A) (B) (C) (D) (E)	56 (A) (B) (C) (D) (E)	81 (A) (B) (C) (D) (E)
7 (A) (B) (C) (D) (E)	32 (A) (B) (C) (D) (E)	57 (A) (B) (C) (D) (E)	82 (A) (B) (C) (D) (E)
8 (A) (B) (C) (D) (E)	33 (A) (B) (C) (D) (E)	58 (A) (B) (C) (D) (E)	83 (A) (B) (C) (D) (E)
9 (A) (B) (C) (D) (E)	34 (A) (B) (C) (D) (E)	59 (A) (B) (C) (D) (E)	84 (A) (B) (C) (D) (E)
10 (A) (B) (C) (D) (E)	35 (A) (B) (C) (D) (E)	60 (A) (B) (C) (D) (E)	85 (A) (B) (C) (D) (E)
11 (A) (B) (C) (D) (E)	36 (A) (B) (C) (D) (E)	61 (A) (B) (C) (D) (E)	86 (A) (B) (C) (D) (E)
12 (A) (B) (C) (D) (E)	37 (A) (B) (C) (D) (E)	62 (A) (B) (C) (D) (E)	87 (A) (B) (C) (D) (E)
13 (A) (B) (C) (D) (E)	38 (A) (B) (C) (D) (E)	63 (A) (B) (C) (D) (E)	88 (A) (B) (C) (D) (E)
14 (A) (B) (C) (D) (E)	39 (A) (B) (C) (D) (E)	64 (A) (B) (C) (D) (E)	89 (A) (B) (C) (D) (E)
15 (A) (B) (C) (D) (E)	40 (A) (B) (C) (D) (E)	65 (A) (B) (C) (D) (E)	90 (A) (B) (C) (D) (E)
16 (A) (B) (C) (D) (E)	41 (A) (B) (C) (D) (E)	66 (A) (B) (C) (D) (E)	91 (A) (B) (C) (D) (E)
17 (A) (B) (C) (D) (E)	42 (A) (B) (C) (D) (E)	67 (A) (B) (C) (D) (E)	92 (A) (B) (C) (D) (E)
18 (A) (B) (C) (D) (E)	43 (A) (B) (C) (D) (E)	68 (A) (B) (C) (D) (E)	93 (A) (B) (C) (D) (E)
19 (A) (B) (C) (D) (E)	44 (A) (B) (C) (D) (E)	69 (A) (B) (C) (D) (E)	94 (A) (B) (C) (D) (E)
20 (A) (B) (C) (D) (E)	45 (A) (B) (C) (D) (E)	70 (A) (B) (C) (D) (E)	95 (A) (B) (C) (D) (E)
21 (A) (B) (C) (D) (E)	46 (A) (B) (C) (D) (E)	71 (A) (B) (C) (D) (E)	96 (A) (B) (C) (D) (E)
22 (A) (B) (C) (D) (E)	47 (A) (B) (C) (D) (E)	72 (A) (B) (C) (D) (E)	97 (A) (B) (C) (D) (E)
23 (A) (B) (C) (D) (E)	48 (A) (B) (C) (D) (E)	73 (A) (B) (C) (D) (E)	98 (A) (B) (C) (D) (E)
24 (A) (B) (C) (D) (E)	49 (A) (B) (C) (D) (E)	74 (A) (B) (C) (D) (E)	99 (A) (B) (C) (D) (E)
25 (A) (B) (C) (D) (E)	50 (A) (B) (C) (D) (E)	75 (A) (B) (C) (D) (E)	100 (A) (B) (C) (D) (E)

Second Day

Second Day Examination

1 According to the Code, the drainage system or any part thereof shall not be covered until:

(A) after a plumbing inspector has been called
(B) it has been tested
(C) it has been inspected
(D) it has been approved

2 According to the Code, drainage systems, when water-tested for tightness and for inspection, shall be tested with no less than a ____ -foot head of water.

(A) 6
(B) 8
(C) 10
(D) 12

3 The Code requires that sewage or other waste from a plumbing system be ____ before it is discharged into any waterway.

(A) discharged through a settling tank
(B) treated with Clorox
(C) rendered innocuous
(D) determined detrimental

4 A single family house connected to a public sewer may discharge all but one of the following substances through its drainage system without violating the Code:

(A) human excrement
(B) potable water
(C) fireplace ashes
(D) clothes washer water

5 According to Code, the ladder required by an inspector so that he can make a final inspection must be furnished by the:

(A) general contractor
(B) roofing contractor
(C) building and zoning department
(D) plumbing contractor

6 The water service pipe supplying water for a two bedroom, one bath house, according to Code, must be:

(A) installed in 20 foot lengths when possible
(B) buried at least 12 inches below grade
(C) no less than ¾" I.D. piping
(D) no less than ¾" O.D. piping

7 According to Code, a single or double sanitary tee may be used in drainage lines where the direction of flow is from:

(A) horizontal to vertical
(B) horizontal to horizontal
(C) vertical to horizontal
(D) vertical to vertical

Second Day Examination

8 The plumber decides to water test the drainage system in its entirety. According to Code, before inspection starts, the water must be kept in the system for a minimum of ____ minutes.

(A) 5   (C) 15
(B) 10   (D) 20

9 Plumbing fixtures, according to Code, cannot be located in such a manner as to:

(A) prevent their use by the physically handicapped
(B) prevent calculation of minimum fixture unit load in sizing drainage piping
(C) confuse the need for additional fixtures
(D) interfere with the normal operation of a door

10 Having a unique plumbing installation, the plumber decides to test the system with air. According to Code, air must be forced into the system until there is a uniform gauge pressure of ____ pounds per square inch.

(A) 5   (C) 15
(B) 10   (D) 20

11 In accordance with Code, horizontally installed piping must be supported at sufficiently close intervals to:

(A) carry the weight of the pipe
(B) carry the weight of the pipe and its contents
(C) prevent strains or stresses
(D) prevent sagging

12 According to Code, bases of soil stacks must be supported:

(A) with concrete blocks
(B) by concrete piers
(C) to the satisfaction of the plumbing inspector
(D) to the satisfaction of the job engineer

13 The Code permits that a plumbing system installed before January 1, 1982, may be altered:

(A) without complying with latest Code requirements
(B) but must comply with latest Code requirements
(C) without obtaining a permit
(D) and Code requirement waived if work is less than 50 percent of the existing system

14 A flushometer valve is a device, according to the Code, which is used to:

(A) flush the contents from a toilet bowl
(B) discharge a predetermined quantity of water to fixtures for flushing purposes
(C) automatically reseal floor drains
(D) flush all surfaces of a floor-mounted urinal

15 The Code defines a vertical vent that is a continuation of the drain to which it connects as a:

(A) common vent
(B) branch vent

(C) relief vent
(D) continuous vent

16 As defined by Code, the primary purpose of a yoke vent is to:

(A) connect a waste stack to a vent stack
(B) connect a vent stack to a soil stack

(C) connect a loop vent to a stack vent
(D) prevent pressure changes in the stacks

17 The primary purpose of the Code in defining terms, rather than words of regular usage, is to:

(A) avoid misunderstanding
(B) include all possible words

(C) make them easier to find
(D) avoid duplication of words

18 The term "liquid waste" as defined by Code is waste that is discharged into a plumbing system from all but one of the following fixtures:

(A) bidet
(B) bed pan washer

(C) wall hung urinal
(D) cuspidor

19 Of the following plumbing terms, the one most nearly correctly defined, according to the Code, is:

(A) "soil pipe" shall mean any pipe manufactured of PVC Schedule 40 which conveys waste to the building drain
(B) "sub-drain" shall mean that portion of a drainage system installed below the first floor level

(C) "indirect waste pipe" shall mean a waste pipe which fails to connect directly to the building drainage system
(D) "stack" shall mean any line of soil, waste or vent piping

20 Of the following plumbing terms, the one most nearly correctly defined, according to the Code, is:

(A) "developed length" shall mean its length along the center line of the pipe and fittings
(B) "continuous waste" shall mean a drain connecting several similar fixtures to one drain pipe

(C) "leader" shall mean any pipe conveying storm water to an approved disposal area
(D) "branch" shall mean any part of the piping system including fittings and risers

21 Back-siphonage is the reverse flow of questionable water from a plumbing fixture into a potable water supply pipe due to _____ in such pipe.

(A) a negative pressure
(B) a positive pressure

(C) a true siphon action
(D) capillary attraction

Second Day Examination

22 The Code defines the depth of a trap seal as being measured from the:

(A) crown weir to the top of the dip of a trap
(B) crown weir to the bottom of the dip of a trap
(C) fixture tailpiece to the crown weir of a trap
(D) fixture tailpiece to the top of the dip of a trap

23 The primary purpose of a vent system, as defined by Code, is a pipe or pipes installed to:

(A) prevent capillary attraction from occurring
(B) prevent negative pressure from developing within the drainage system
(C) prevent positive pressure from being created within the drainage system
(D) provide a circulation of air within the drainage system

24 A Durham System, as defined by Code, is a system where:

(A) piping and fittings are welded
(B) piping and fittings are of the flare type
(C) piping is threaded and fittings are of the recessed type
(D) piping and fittings are of the flanged type

25 Of the following piping installations, the one that is not defined by Code as a "stack" when installed as a vertical main is:

(A) soil piping
(B) water piping
(C) waste piping
(D) vent piping

26 The Code requires that sheet lead used in the construction of safe pans for shower compartments weigh no less than _____ pounds per square foot.

(A) 2
(B) 2½
(C) 3
(D) 4

27 According to Code, seamless copper water tubing used for water distribution must meet the _____ standards.

(A) ASTM
(B) ANSI
(C) PS
(D) AWWA

28 According to Code, the material used in the manufacture of cast-brass flared fittings is classified as being:

(A) ferrous
(B) nonmetallic
(C) nonferrous
(D) metallic

Second Day Examination

29 Caulked-on floor flanges for water closets, according to Code, must have a minimum caulking depth of ___ inch(es).

(A) 1

(B) 1½

(C) 2

(D) 2½

30 Of the following listed standards, the one that is <u>not</u> a Code-adopted standard for cast-iron pipe and fitting is:

(A) ANSI

(B) ASME

(C) ASTM

(D) FS

31 The Administrative Authority may approve alternate materials that are not specifically defined in the Code, provided it is clear that:

(A) the Code is not applicable

(B) minimum standards are equal

(C) the manufacturer's instructions for installation are adhered to

(D) the manufacturer will assume full responsibility for his material

32 Acrylonitrile-butadiene-styrene piping and fittings, according to Code, would most generally be used for:

(A) above-ground water piping

(B) swimming pool pressure piping

(C) drainage systems

(D) natural gas yard piping

33 The material used in the manufacture of galvanized steel piping, according to Code, is classified as being:

(A) ferrous

(B) nonferrous

(C) carbon

(D) silicon and manganese

34 The wall thickness of lead bends used in a plumbing system, according to Code, must be ___ inch thick.

(A) 1/16

(B) 1/8

(C) 3/16

(D) 1/4

35 Code mandates that joints and connections in the plumbing system be gastight and watertight for:

(A) the intended use

(B) the working pressure available

(C) the pressures required by test

(D) any fluctuation of pressure

36 The Code will permit the installation of elastomeric compression-type joints as an alternate for _____ on hub and spigot cast-iron soil pipe.

(A) lead and oakum joints

(B) stainless steel no-hub couplings

(C) cement and oakum joints

(D) hot poured compound joints

37 Of the following materials, the one that is used to make up a threaded joint between copper tubing and galvanized steel piping is a:

(A) brass male adapter

(B) dresser coupling

(C) approved kafer fitting

(D) galvanized extension piece

38 Of the differing types of joints listed below, the one which is Code-approved for lead pipe is the:

(A) sisson joint

(B) burned joint

(C) wiped joint

(D) soldered joint

39 According to the Code, cleanouts on the seal of a trap are prohibited on:

(A) residential lavatory traps

(B) commercial sink traps

(C) stall shower traps

(D) barber shop sink traps

40 According to the Code, 3-inch cleanouts **must** be accessible and have a minimum clearance for cleaning purposes of no less than _____ inches.

(A) 6

(B) 12

(C) 18

(D) 24

41 According to the Code, the integrity of a trap seal is protected by:

(A) proper venting design

(B) having a trap seal of more than four inches

(C) having slip joints on both sides of the trap seal

(D) having slip joints on the inlet side of the trap seal only

42 The Code requires that a cleanout be installed on each 4-inch horizontal drainage pipe:

(A) at the downstream end of the pipe

(B) at the upstream end of the pipe

(C) no more than 50 feet apart

(D) no less than 100 feet apart

43 According to Code, all of the following types of traps are prohibited for use in a drainage system, **except**:

(A) the bell trap

(B) the crown-vented trap

(C) the P-trap

(D) the full "S" trap

Second Day Examination

44 Where the Administrative Authority allows manholes to be installed on a building sewer, Code requires that such manholes be no less than _____ feet apart.

(A) 100

(B) 150

(C) 200

(D) 300

45 According to Code, of the approved traps listed below, the one most successful in staying clogfree is the:

(A) "S" trap

(B) trap having movable parts

(C) Bell trap

(D) P-trap

46 Fittings used in a PVC type DWV drainage system shall:

(A) not be used where combustible construction is allowed

(B) conform to the type of pipe used

(C) be Schedule 80 type only

(D) be of the recessed type

47 To meet Code requirements, a 2-inch diameter waste pipe installed horizontally must fall a minimum of _____ inches every 35 feet.

(A) 5-1/8

(B) 6-5/8

(C) 7-1/4

(D) 8-3/4

48 According to the Code, _____ must discharge their waste through a grease interceptor.

(A) garbage can washers

(B) commercial food grinders

(C) garage floor drains

(D) acid waste systems

49 A 4-inch diameter building drain installed horizontally must have a minimum of _____-inch fall per foot to meet the Code.

(A) 1/16

(B) 1/8

(C) 1/4

(D) 1/2

50 Of the following drainage piping materials used within a building, the one prohibited by Code for underground installation is:

(A) DWV copper piping

(B) lead piping

(C) steel piping

(D) brass piping

51 Of the following factors, the one that is not Code-required to use for determining the size of a drainage pipe is:

(A) the grade of the piping

(B) the type of piping material used

(C) the type of fixtures used

(D) the length of run

Second Day Examination

52 The minimum sizes of horizontal drainage piping, according to Code, are determined by:

(A) the waste outlet size of fixtures connected thereto

(B) the number and type of fixtures connected thereto

(C) the length of run and pitch per foot

(D) the total of all fixture units connected thereto

53 In the case of vertical drainage piping, the minimum sizes, according to Code, are determined by the total of all fixture units connected thereto, and by:

(A) any offsets in the stack

(B) their length

(C) the minimum vent piping size required

(D) the largest fixture opening at the highest level

54 Where there is a continuous flow from an air conditioning unit into a building drainage system, each gpm of flow, according to Code, must equal _____ fixture unit(s).

(A) ½

(B) 1

(C) 1½

(D) 2

55 The Code generally requires that piping with a high percentage of silicon be installed to convey:

(A) backwash waste from swimming pools

(B) waste containing acid

(C) waste from industrial plants

(D) waste from hospital operating rooms

56 Code requires that when subsoil drainage systems are installed, they must discharge:

(A) into an approved sump

(B) through an approved oil and sand interceptor

(C) through bucket-type floor drains

(D) in a manner satisfactory to the plumbing inspector

57 Given: a restaurant serving 150 people at one seating. Requirement: a full size grease interceptor. Of the following, the _____ is prohibited by Code from discharging into the greasy waste system.

(A) garbage can washer

(B) three-compartment pot sink

(C) commercial dishwasher

(D) food-waste grinder

58 Where frost or snow closure is likely to occur, each vent extension through a roof must be at least 3 inches in diameter. Code specifies that for a 2-inch vent, the change in diameter must be made inside the building at least _____ inches below the roof.

(A) 6

(B) 12

(C) 18

(D) 24

59 According to Code, a relief yoke vent may be installed:

(A) below the fixture branch serving that floor

(B) above the fixture branch serving that floor

(C) at the same level as the fixture branch serving that floor

(D) above the flow line of the fixture branch serving that floor

60 A vent pipe sometimes has to be run horizontally to avoid a pass-through window installed above a kitchen sink. This may be done, according to Code, if the horizontal vent pipe is at least _____ inches above the flood-level rim of the sink.

(A) 2
(B) 4

(C) 6
(D) 8

61 Sumps receiving waste from plumbing fixtures, according to Code, shall be vented with a minimum size vent of _____ inches.

(A) 1¼
(B) 1½

(C) 2
(D) 3

62 Code permits an oil interceptor which collects liquid waste from a garage floor to discharge its waste after treatment directly into a:

(A) catch basin
(B) building sewer

(C) building storm sewer
(D) properly sized dry well

63 In designing a horizontal waste and vent system, the Code places a maximum drop between the fixture and its trap. This distance must not exceed _____ inches.

(A) 12
(B) 18

(C) 20
(D) 24

64 The liquid capacity of all septic tanks for dwelling units, the Code specifies, is determined by the number:

(A) of bedrooms
(B) of persons

(C) of bathrooms
(D) and type of plumbing fixtures

65 Where leaching beds are permitted in lieu of trenches, the Code specifies that the area of each such bed must be:

(A) 25 percent greater
(B) 25 percent less

(C) 50 percent greater
(D) 50 percent less

66   Given: A cesspool has been approved by the Administrative Authority as a sewage disposal facility for a single family residence. The source of water is a domestic well. The soil criteria are clay with considerable amounts of sand and gravel. The minimum horizontal distance between the cesspool and the well, according to Code, must not be less than _____ feet.

(A) 100
(B) 125

(C) 150
(D) 175

67   The one of the following uses not requiring potable water (as defined by Code) is water used for:

(A) land irrigation
(B) culinary purposes

(C) domestic households
(D) private swimming pools

68   Water pressure in public water mains averages approximately 45 to 50 pounds per square inch. The Code considers that pressure in public water mains is often excessive and requires that a pressure-reducing valve be installed when the pressure exceeds _____ pounds per square inch.

(A) 65
(B) 70

(C) 75
(D) 80

69   According to Code, if the highest group of fixtures contains flushometer valves, the minimum pressure required for the fixtures to function properly would be _____ psi.

(A) 8
(B) 11

(C) 15
(D) 18

70   To meet Code, a three-story apartment building having tank-type flush water closets must have a minimum pressure of _____ psi on the third floor.

(A) 8
(B) 11

(C) 15
(D) 18

71   A lawn sprinkler system receives its source of water from the building supply line. The Code requires the installation of a vacuum breaker to prevent reverse flow into the potable water supply. According to Code, such vacuum breaker must be installed:

(A) in an accessible valve box

(B) at least 6 inches above the highest sprinkler head

(C) a minimum of 4 inches below the highest sprinkler head

(D) to isolate the sprinkler system from other water sources

Second Day Examination

72 A 10-foot section of cold water piping is unavoidably installed in the exterior wall of an apartment building in Crossville, Tennessee. The plumbing inspector turns down the plumber's work because he:

(A) used a pipe size larger than required by Code

(B) had not graded the pipe to drain dry

(C) failed to allow for contraction and expansion of the pipe

(D) did not make adequate provisions to protect it from freezing

73 The water pipe sizing procedure, according to Code, is based on:

(A) static pressure loss or gain

(B) pressure required at fixtures to produce required flow

(C) the friction or pressure loss through a water meter

(D) a system of pressure requirements and losses

74 After cutting the desired length of type L copper water tubing to fit between an elbow and a tee, the Code requires that:

(A) the outside be cleaned bright

(B) the joints and pipe be properly fluxed

(C) a 50-50 solder be used to make the joints

(D) the tubing be returned to full bore

75 In sizing roof drains and storm drainage piping, _____ percent of a vertical wall which diverts rain water to the roof must be added to the projected roof area when calculating the Code-required storm drainage piping.

(A) 50

(B) 25

(C) 40

(D) 60

76 Given: A job calls for two roof deck strainers on a sun deck. The Code would require that such drains have a minimum inlet area _____ times the area of the pipe to which the drain is connected.

(A) 1

(B) 1½

(C) 2

(D) 2½

77 Because of the limited fall available for the installation of a building storm drainage system, a fall of 1/8 inch per foot is used. If the maximum rainfall is 4 inches per hour and the roof area is 27,350 square feet, the Code-required minimum pipe size for the building storm sewer is _____ inches.

(A) 8

(B) 10

(C) 12

(D) 15

Second Day Examination

78 Which of the following statements is most nearly correct, according to Code?

(A) The minimum grade for horizontal rain water piping acceptable by Code is 1/16-inch fall per foot

(B) The minimum size rain leader acceptable by Code is 2 inches in diameter

(C) All Codes base their storm drainage pipe sizes on 4 inches of rainfall per hour

(D) When sizing rain leaders, projected roof areas need not be considered

79 A 30-gallon electric water heater is installed under a staircase in a two-story residence. The relief valve drain line, according to Code, must extend to the exterior of the building and terminate a minimum of _____ inches above grade.

(A) 6
(B) 8

(C) 10
(D) 12

80 A 40-gallon water heater requires a combination temperature and pressure relief valve. The valve drain outlet is ½-inch. According to Code, the drainline, then, should be _____ -inch.

(A) 1/4
(B) 3/8

(C) 1/2
(D) 5/8

81 The one of the following which is <u>prohibited</u> by Code for restaurant use is the:

(A) siphon jet water closet
(B) washout type water closet

(C) reverse trap water closet
(D) siphon action water closet

82 Many Codes consider floor drains as plumbing fixtures. According to Code, which of the following statements is most nearly correct?

(A) It is not necessary to install floor drains in public toilet rooms

(B) The Code mandates the installation of floor drains in public toilet rooms

(C) The drain inlet must be located in full view

(D) Floor drains must be located where strainers may not cause a hazard

83 Given: You are replacing a galvanized steel pipe lavatory drain with copper pipe. The connection to the cast-iron stack must be made with a solder bushing. Code requires that the minimum weight of the solder bushing be _____ ounces.

(A) 4
(B) 6

(C) 8
(D) 10

Second Day Examination

84 According to Code, when installing wall-hung water closets,

(A) the bottom of the water closet flange must be set on an approved firm base

(B) the water closet bowl must be of the elongated type

(C) the water closet must be securely bolted to an approved carrier fitting

(D) the closet bend must be cut off so as to present a smooth surface

85 When it becomes necessary to connect any additional gas appliance(s) to an existing gas system, the Code requires verification that:

(A) existing gas piping is properly supported

(B) existing gas piping has adequate capacity

(C) a master plumber supervises such work

(D) existing gas piping is in good physical condition

86 A gas company's employees are installing a new service line and meter for a dry cleaning establishment. According to Code,

(A) the work must be inspected by the plumbing official

(B) the gas company shall first obtain a permit

(C) the gas company does not need a permit

(D) the installation of the service line and meter shall be in accordance with approved plans

87 Some gases are considered to be corrosive. Where such gases are used, the Code mandates that _____ cannot be used for consumers' gas piping.

(A) wrought iron pipe

(B) wrought steel pipe

(C) brass pipe in iron pipe sizes

(D) galvanized steel pipe

88 When it becomes necessary to connect plastic pipe to metallic pipe in a gas system, Code requires that the connection be made:

(A) outside of the building, underground

(B) outside of the building, above ground

(C) with compression type mechanical joints only

(D) with wrought iron pipe and fittings only

89 A pipe which conveys gas from a supply line to the appliance is defined by Code as a:

(A) branch line

(B) dead end

(C) gas main

(D) house line

90 Air that is supplied to the flame at the point of combustion is defined by Code as:

(A) secondary air

(B) primary air

(C) flue air

(D) combustion air

Second Day Examination

91 According to Code, when installing wrought steel gas piping within a building, the minimum size for pipe outlets must be _____ -inch.

(A) 1/4   (C) 1/2
(B) 3/8   (D) 3/4

92 Code states that immediately after installation, each outlet must be securely closed, gastight, and must be left closed until:

(A) all testing has been completed   (C) an appliance is connected to the outlet
(B) gas lines have been purged   (D) accepted by the gas inspector

93 The Code requirement for a hot water heater with electric ignition system is that it must ignite the pilot within _____ seconds after the gas supply is turned on.

(A) 5   (C) 15
(B) 10   (D) 20

94 According to Code, a central gas-fired heating boiler utilizing complete shut-off type safety devices must be provided with a _____ ahead of all controls.

(A) manual shut-off valve   (C) pressure gauge
(B) drip valve   (D) relief valve

95 According to Code, exit terminals of mechanical draft systems must be located a minimum of _____ feet above grade when adjacent to public walkways.

(A) 6   (C) 8
(B) 7   (D) 9

96 Of the following occupancies, the one classified by Code as "light hazard" is:

(A) machine shops   (C) feed mills
(B) mercantiles   (D) museums

97 The minimum base diameter of an installed fire standpipe in a 16-story office building, according to Code, cannot be smaller than _____ inches.

(A) 4   (C) 6
(B) 5   (D) 8

Second Day Examination

98 According to the Code, an unlined fire hose shall be provided with a:

(A) soft-seat check valve

(B) solenoid valve

(C) listed automatic drip connection

(D) NFPA approved fire truck connector

99 The gas content of a medical gas piping system must be readily identifiable by labeling at intervals not to exceed _____ feet, according to Code.

(A) 8

(B) 10

(C) 15

(D) 20

100 In accordance with Code, the plumber must be certain that all test gas is removed from any newly-installed medical gas piping system by making sure the gas:

(A) flows through and out each outlet until it has no blue haze left

(B) flows through and out each outlet until a flame will no longer ignite the gas

(C) no longer discolors a white cloth which it impinges on

(D) flows freely from each outlet, giving off no odor

First Day Examination Answer Sheet
(Morning Portion)

Question number	Correct answer	Study question	Question number	Correct answer	Study question	Question number	Correct answer	Study question
1	C	15-2	8	C	17-43	15	B	13-22
2	A	14-7	9	B	20-3	16	D	3-18
3	B	14-3	10	D	20-21	17	B	11-17
4	C	14-17	11	D	20-25	18	A	20-23
5	C	16-3	12	C	11-4			
6	C	16-36	13	A	11-33	19	D	2-46
7	B	17-40	14	B	13-4	20	A	20-4

First Day Examination Answer Sheet
(Afternoon Portion)

Question number	Correct answer	Study question	Question number	Correct answer	Study question	Question number	Correct answer	Study question
1	B	1-4	31	D	9-7	**Zoned Drawings**		
2	C	1-9	32	C	9-29			
3	B	1-19	33	B	10-2	61	C	6-34
4	A	1-23	34	C	10-4	62	B	7-1
5	C	1-26	35	A	10-16	63	A	7-17
						64	D	1-33
6	D	1-29	36	C	11-19	65	B	Refer to Code
7	C	1-36	37	B	11-22			
8	D	1-48	38	D	11-26			
9	A	2-2	39	B	12-3	66	D	9-6
10	A	2-5	40	C	12-6	67	C	Refer to Code
11	A	2-52	41	B	12-8	68	A	Refer to Code
12	C	2-54	42	B	12-14			
13	D	3-2	43	A	12-15	69	B	7-21
14	D	3-8	44	D	13-11	70	D	7-22
15	C	3-10	45	B	13-20			
16	A	3-14	46	D	13-21			
17	A	3-19	47	C	13-27	**Mini Drawings**		
18	D	3-22	48	B	13-33	**(See corrected drawings in Answer section beginning on page 250)**		
19	B	5-1	49	C	13-37			
20	B	5-17	50	A	13-39			
21	D	6-9	51	C	20-1	71	A	17-43
22	B	6-12	52	C	20-2	72	A	16-34
23	C	6-16	53	D	20-6	73	A	16-13
24	C	6-21	54	D	20-8	74	B	19-1
25	C	6-28	55	B	20-10	75	B	19-8
26	B	7-6	56	C	20-11	76	A	Refer to Code
27	D	7-20	57	C	20-15	77	B	19-13
28	A	8-2	58	C	20-19	78	A	7-11
29	C	8-15	59	C	20-33	79	B	19-10
30	B	8-19	60	C	20-26	80	B	19-11

Second Day Examination
Answer Sheet

Question number	Correct answer	Study question	Question number	Correct answer	Study question	Question number	Correct answer	Study question
1	D	1-3	34	B	3-21	67	A	11-20
2	C	1-6	35	C	4-1	68	D	11-21
3	C	1-10				69	C	11-23
4	C	1-13	36	A	4-5	70	A	11-24
5	D	1-16	37	A	4-9			
			38	C	4-10	71	B	11-25
6	C	1-20	39	C	5-2	72	D	11-27
7	A	1-24	40	C	5-4	73	D	11-29
8	C	1-28				74	D	11-30
9	D	1-39	41	A	5-6	75	A	12-4
10	A	1-41	42	B	5-10			
			43	C	5-14	76	C	12-10
11	D	1-43	44	D	5-33	77	C	12-11
12	C	1-46	45	D	5-35	78	B	12-13
13	A	1-52				79	A	13-8
14	B	2-1	46	B	6-6	80	C	13-9
15	D	2-4	47	D	6-7			
			48	A	6-8	81	B	13-10
16	D	2-6	49	B	6-11	82	C	13-14
17	A	2-8	50	C	6-13	83	B	13-16
18	B	2-10				84	C	13-17
19	C	2-24	51	B	6-17	85	B	14-4
20	A	2-25	52	D	6-18			
			53	B	6-19	86	C	14-9
21	A	2-30	54	D	6-20	87	C	14-15
22	A	2-45	55	B	6-27	88	A	14-29
23	D	2-48				89	A	14-35
24	C	2-50	56	A	6-36	90	A	14-36
25	B	2-55	57	D	6-38			
			58	B	7-8	91	C	15-15
26	D	3-1	59	A	7-12	92	C	15-33
27	A	3-4	60	C	7-18	93	C	16-6
28	C	3-5				94	A	16-10
29	C	3-7	61	B	7-23	95	B	16-33
30	B	3-11	62	B	8-1			
			63	D	9-30	96	D	17-5
31	B	3-13	64	A	10-1	97	C	17-7
32	C	3-15	65	C	10-3	98	C	17-13
33	A	3-17	66	C	10-15	99	D	18-10
						100	C	18-9

Index

317

Other Practical References

Estimating Plumbing Costs

Offers a basic procedure for estimating materials, labor, and direct and indirect costs for residential and commercial plumbing jobs. Explains how to interpret and understand plot plans, design drainage, waste, and vent systems, meet code requirements, and make an accurate take-off for materials and labor. Includes sample cost sheets, manhour production tables, complete illustrations, and all the practical information you need to accurately estimate plumbing costs. **224 pages, 8½ x 11, $17.25**

Plumbers Handbook Revised

This new edition shows what will and what will not pass inspection in drainage, vent, and waste piping, septic tanks, water supply, fire protection, and gas piping systems. All tables, standards, and specifications are completely up-to-date with recent changes in the plumbing code. Covers common layouts for residential work, how to size piping, selecting and hanging fixtures, practical recommendations and trade tips. This book is the approved reference for the plumbing contractors exam in many states. **240 pages, 8½ x 11, $18.00**

Planning and Designing Plumbing Systems

Explains in clear language, with detailed illustrations, basic drafting principles for plumbing construction needs. Covers basic drafting fundamentals: isometric pipe drawing, sectional drawings and details, how to use a plot plan, and how to convert it into a working drawing. Gives instructions and examples for water supply systems, drainage and venting, pipe, valves and fixtures, and has a special section covering heating systems, refrigeration, gas, oil, and compressed air piping, storm, roof and building drains, fire hydrants, and more. **224 pages, 8½ x 11, $13.00**

Planning Drain, Waste, and Vent Systems

Anyone who has designed plumbing systems knows from experience how important it is to follow the code exactly. Even a small oversight can keep a plan from being approved and cause expensive delays. Unfortunately, the plumbing code isn't easy to understand, so it's easy to make mistakes. In this new book, Howard Massey, a recognized expert on the plumbing codes and author of several plumbing manuals, shows you how to design drainage, waste and vent systems so you never lose time and money by having your plans rejected by the examiner or engineer. **192 pages, 8½ x 11, $19.25**

Audiotape: Plumber's Exam

These tapes are made to order for the busy plumber looking for a better paying career as a licensed apprentice, journeyman or master plumber. Howard Massey, who developed the tapes, has written many of the questions used on plumber's exams, and has monitored and graded the exam. He knows what you need to pass. This two-audiotape set asks you over 100 often-used exam questions in an easy-to-remember format. This is the easiest way to study for the exam. **Two 60 minute audio tapes, $19.95**

Manual of Professional Remodeling

This is the practical manual of professional remodeling written by an experienced and successful remodeling contractor. Shows how to evaluate a job and avoid 30-minute jobs that take all day, what to fix and what to leave alone, and what to watch for in dealing with subcontractors. Includes chapters on calculating space requirements, repairing structural defects, remodeling kitchens, baths, walls and ceilings, doors and windows, floors, roofs, installing fireplaces and chimneys (including built-ins), skylights, and exterior siding. Includes blank forms, checklists, sample contracts, and proposals you can copy and use. **400 pages, 8½ x 11, $19.75**

Paint Contractor's Manual

How to start and run a profitable paint contracting company: getting set up and organized to handle volume work, avoiding the mistakes most painters make, getting top production from your crews and the most value from your advertising dollar. Shows how to estimate all prep and painting. Loaded with manhour estimates, sample forms, contracts, charts, tables and examples you can use. **224 pages, 8½ x 11, $19.25**

National Construction Estimator

Current building costs in dollars and cents for residential, commercial and industrial construction. Prices for every commonly used building material, and the proper labor cost associated with installation of the material. Everything figured out to give you the "in place" cost in seconds. Many time-saving rules of thumb, waste and coverage factors and estimating tables are included. **544 pages, 8½ x 11, $19.50. Revised annually.**

Contractor's Survival Manual

How to survive hard times in construction and take full advantage of the profitable cycles. Shows what to do when the bills can't be paid, finding money and buying time, transferring debt, and all the alternatives to bankruptcy. Explains how to build profits, avoid problems in zoning and permits, taxes, time-keeping, and payroll. Unconventional advice includes how to invest in inflation, get high appraisals, trade and postpone income, and how to stay hip-deep in profitable work. **160 pages, 8½ x 11, $16.75**

Residential Wiring

Shows how to install rough and finish wiring in both new construction and alterations and additions. Complete instructions are included on troubleshooting and repairs. Every subject is referenced to the 1987 National Electrical Code, and over 24 pages of the most needed NEC tables are included to help you avoid errors so your wiring passes inspection — the first time. **352 pages, 5½ x 8½, $18.25**

Wood Frame House Construction

From the layout of the outer walls, excavation and formwork, to finish carpentry, and painting, every step of construction is covered in detail with clear illustrations and explanations. Everything the builder needs to know about framing, roofing, siding, insulation and vapor barrier, interior finishing, floor coverings, and stairs ... complete step by step "how to" information on what goes into building a frame house. **240 pages, 8½ x 11, $14.25. Revised edition.**

Bookkeeping for Builders

This book will show you simple, practical instructions for setting up and keeping accurate records — with a minimum of effort and frustration. Shows how to set up the essentials of a record keeping system: the payment journal, income journal, general journal, records for fixed assets, accounts receivable, payables and purchases, petty cash, and job costs. You'll be able to keep the records required by the I.R.S., as well as accurate and organized business records for your own use. **208 pages, 8½ x 11, $19.75**

Carpentry Estimating

Simple, clear instructions show you how to take off quantities and figure costs for all rough and finish carpentry. Shows how much overhead and profit to include, how to convert piece prices to MBF prices or linear foot prices, and how to use the tables included to quickly estimate manhours. All carpentry is covered: floor joists, exterior and interior walls and finishes, ceiling joists and rafters, stairs, trim, windows, doors, and much more. Includes sample forms, checklists, and the author's factor worksheets to save you time and help prevent errors. **320 pages, 8½ x 11, $25.50**

Cost Records for Construction Estimating

How to organize and use cost information from jobs just completed to make more accurate estimates in the future. Explains how to keep the cost records you need to reflect the time spent on each part of the job. Shows the best way to track costs for sitework, footing, foundations, framing, interior finish, siding and trim, masonry, and subcontract expense. Provides sample forms. **208 pages, 8½ x 11, $15.75**

Contractor's Guide to the Building Code Rev.

This completely revised edition explains in plain English exactly what the Uniform Building Code requires and shows how to design and construct residential and light commercial buildings that will pass inspection the first time. Suggests how to work with the inspector to minimize construction costs, what common building short cuts are likely to be cited, and where exceptions are granted. **544 pages, 5½ x 8½, $24.25**

Builder's Office Manual, Revised

Explains how to create routine ways of doing all the things that must be done in every construction office — in the minimum time, at the lowest cost, and with the least supervision possible: Organizing the office space, establishing effective procedures and forms, setting priorities and goals, finding and keeping an effective staff, getting the most from your record-keeping system (whether manual or computerized). Loaded with practical tips, charts and sample forms for your use. **192 pages, 8½ x 11, $15.50**

Basic Plumbing with Illustrations

The journeyman's and apprentice's guide to installing plumbing, piping and fixtures in residential and light commercial buildings: how to select the right materials, lay out the job and do professional quality plumbing work. Explains the use of essential tools and materials, how to make repairs, maintain plumbing systems, install fixtures and add to existing systems. **320 pages, 8½ x 11, $22.00**

Builder's Guide to Accounting Revised

Step-by-step, easy to follow guidelines for setting up and maintaining an efficient record keeping system for your building business. Not a book of theory, this practical, newly-revised guide to all accounting methods shows how to meet state and federal accounting requirements, including new depreciation rules, and explains what the tax reform act of 1986 can mean to your business. Full of charts, diagrams, blank forms, simple directions and examples. **304 pages, 8½ x 11, $17.25**

Construction Estimating Reference Data

Collected in this single volume are the building estimator's 300 most useful estimating reference tables. Labor requirements for nearly every type of construction are included: site work, concrete work, masonry, steel, carpentry, thermal & moisture protection, doors and windows, finishes, mechanical and electrical. Each section explains in detail the work being estimated and gives the appropriate crew size and equipment needed. **368 pages, 11 x 8½, $26.00**

Electrical Construction Estimator

If you estimate electrical jobs, this is your guide to current material costs, reliable manhour estimates per unit, and the total installed cost for all common electrical work: conduit, wire, boxes, fixtures, switches, outlets, loadcenters, panelboards, raceway, duct, signal systems, and more. Explains what every estimator should know before estimating each part of an electrical system. **416 pages, 8½ x 11, $25.00. Revised annually**

Rough Carpentry

All rough carpentry is covered in detail: sills, girders, columns, joists, sheathing, ceiling, roof and wall framing, roof trusses, dormers, bay windows, furring and grounds, stairs and insulation. Many of the 24 chapters explain practical code approved methods for saving lumber and time without sacrificing quality. Chapters on columns, headers, rafters, joists and girders show how to use simple engineering principles to select the right lumber dimension for whatever species and grade you are using. **288 pages, 8½ x 11, $17.00**

Estimating Home Building Costs

Estimate every phase of residential construction from site costs to the profit margin you should include in your bid. Shows how to keep track of manhours and make accurate labor cost estimates for footings, foundations, framing and sheathing finishes, electrical, plumbing and more. Explains the work being estimated and provides sample cost estimate worksheets with complete instructions for each job phase. **320 pages, 5½ x 8½, $17.00**

BUSINESS REPLY MAIL
FIRST CLASS   PERMIT NO. 271   CARLSBAD, CA

POSTAGE WILL BE PAID BY ADDRESSEE

Craftsman Book Company
6058 Corte Del Cedro
P. O. Box 6500
Carlsbad, CA 92008-0992

NO POSTAGE
NECESSARY
IF MAILED
IN THE
UNITED STATES

BUSINESS REPLY MAIL
FIRST CLASS   PERMIT NO. 271   CARLSBAD, CA

POSTAGE WILL BE PAID BY ADDRESSEE

Craftsman Book Company
6058 Corte Del Cedro
P. O. Box 6500
Carlsbad, CA 92008-0992

NO POSTAGE
NECESSARY
IF MAILED
IN THE
UNITED STATES

BUSINESS REPLY MAIL
FIRST CLASS   PERMIT NO. 271   CARLSBAD, CA

POSTAGE WILL BE PAID BY ADDRESSEE

Craftsman Book Company
6058 Corte Del Cedro
P. O. Box 6500
Carlsbad, CA 92008-0992

NO POSTAGE
NECESSARY
IF MAILED
IN THE
UNITED STATES